Joseph D. Fehribach
Sequences and Series in Calculus

Also of Interest

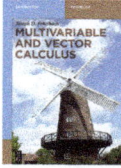

Multivariable and Vector Calculus
Joseph D. Fehribach, 2020
ISBN 978-3-11-066020-3, e-ISBN 978-3-11-066060-9

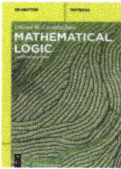

Mathematical Logic
An Introduction
Daniel Cunningham, 2023
ISBN 978-3-11-078201-1, e-ISBN 978-3-11-078219-6

Abstract Algebra
An Introduction with Applications
Derek J. S. Robinson, 2022
ISBN 978-3-11-068610-4, e-ISBN 978-3-11-069116-0

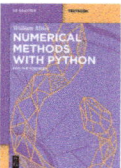

Numerical Methods with Python
for the Sciences
William Miles, 2023
ISBN 978-3-11-077645-4, e-ISBN 978-3-11-077664-5

Analysis with Mathematica®
Galina Filipuk, Andrzej Kozłowski
Volume 1 Single Variable Calculus, 2019
ISBN 978-3-11-059013-5, e-ISBN 978-3-11-059014-2
Volume 2 Multi-variable Calculus, 2020
ISBN 978-3-11-066038-8, e-ISBN 978-3-11-066039-5
Volume 3 Differential Geometry, Differential Equations, and Special Functions, 2022
ISBN 978-3-11-077454-2, e-ISBN 978-3-11-077464-1

Joseph D. Fehribach

Sequences and Series in Calculus

—

DE GRUYTER

Mathematics Subject Classification 2020
Primary: 26A06; Secondary: 26A09

Author
Prof. Dr. Joseph D. Fehribach
Worcester Polytechnic Institute
Dept. of Mathematical Sciences
100 Institute Road
Worcester MA 01609-2280
United States of America
bach@math.wpi.edu

ISBN 978-3-11-076835-0
e-ISBN (PDF) 978-3-11-076839-8
e-ISBN (EPUB) 978-3-11-076846-6

Library of Congress Control Number: 2023933768

Bibliographic information published by the Deutsche Nationalbibliothek
The Deutsche Nationalbibliothek lists this publication in the Deutsche Nationalbibliografie;
detailed bibliographic data are available on the Internet at http://dnb.dnb.de.

© 2023 Walter de Gruyter GmbH, Berlin/Boston
Cover image: Mary Baratto / iStock / Getty Images Plus
Typesetting: VTeX UAB, Lithuania
Printing and binding: CPI books GmbH, Leck

www.degruyter.com

To my family—Jan, Jamie, Jess and Andrew

Augustin-Louis Cauchy (1789–1857) whose work is the basis for many of the theorems in this text. (This work is in the public domain (PD-US). https://commons.wikimedia.org/wiki/File:Augustin_Louis_ Cauchy.jpg)

Foreword

This text is a kind of prequel to my first book *Multivariable and Vector Calculus*. For the past eight years or so, I have taught a two-course sequence here at WPI intended for our entering first-year students who come to WPI ready to study first sequences and series in calculus and then multivariable and vector calculus, and who are interested in a more mathematical, proof-oriented treatment of this material. After finishing the first book in 2020, I was offered the opportunity to write this second text as well, and I was happy to do that.

As the title suggests, this book covers how sequences and series can be used in calculus. As with my first book, it is written, I hope, at a relatively high level, designed for students who have earned high marks on the AP Calculus AB exam (American system) or a maths A-level (British system), and who are interested in learning about these topics in some depth. This book is an alternative to the standard 1000-page calculus text that covers all of classical calculus with many applications and much discussion of ancillary material. Because this text is relatively brief, there is no remedial material, little discussion of computing, and no extensive treatment of applications (though a number of connections to science or engineering are highlighted). What is here is The Calculus and analysis, with proofs, as things stood roughly at the time of Cauchy. There is no discussion of compactness or metric spaces, though some of the work of Cantor does appear in Chapter 2. My hope is that its brevity will make it more accessible to students. It can be used as the text for a course or for self-study by students working on their own. But in any case, I hope that it is useful to students trying to master sequences and series in calculus.

This book assumes that the reader has studied single-variable calculus at the level of AP Calculus AB. As such, it assumes that readers are familiar with, among other things, how to carry out basic integration via substitution. It does not assume that readers necessarily know several somewhat more advanced techniques such as integration by parts and partial fraction integration. These topics are discussed and reviewed explicitly in Chapter 5.

Worcester, Massachusetts   Joseph D. Fehribach
February 2023

https://doi.org/10.1515/9783110768398-201

Notes to the reader

Readers should be aware of the following points as they work through this book:
- Throughout the book, several bits of mathematical shorthand are included, in part to help the reader understand what they mean and how they are used.
  - $\in$: is in, is an element of
  - $\subset$: is contained in, is a subset of
  - **iff**: if and only if
  - **DNE**: does not exist
  - $\simeq$: is approximately equal to
  - $\to$: approaches, goes to
  - $\Longrightarrow$, $\Rightarrow$: implies
  - $\Longleftarrow$, $\Leftarrow$: is implied by, whenever
  - $\Longleftrightarrow$, $\Leftrightarrow$: is equivalent to (logically the same as iff)
  - $\lceil\ \rceil$: ceiling function, the next greater integer
  - $\forall$: for all, for every
  - $\exists$: there exists; for some (when the symbol starts a prepositional phrase)
  - $\mathbb{Z}$: the integers (*Zahlen* in German)
  - $\mathbb{Z}^+$: the positive integers
  - $\mathbb{Q}$: the rational (quotient) numbers
  - $\mathbb{R}$: the real numbers
- There are three symbols for equality used here:
  - The symbol ":=" is read as "defined equal to" and is used in definitions to indicate that the new entity on the left of the colon is defined to be the previously discussed entity on the right.
  - The symbol "$\equiv$" is read as "identically equal to" and means that the two equalities on either side are always equal; so, for example, $\sin^2\theta + \cos^2\theta \equiv 1$ no matter what angle θ is chosen.
  - The standard equal sign "=" is used in all other cases; so $x = 2$ in one example, but x may take on other values in other discussions.
- If f is a real-valued function defined on a domain D that is a subset of $\mathbb{R}$, this arrangement is expressed in symbols as $f : D \subset \mathbb{R} \to \mathbb{R}$. If f is continuous on its domain, this is symbolized as $f \in C(D)$. If f is n times differentiable and its n-th derivative $f^{(n)}$ is itself continuous on its domain, then one writes $f \in C^n(D)$.

https://doi.org/10.1515/9783110768398-202

Contents

1 Sequences in $\mathbb{R}$

Our discussion begins with a very basic concept in mathematics: sequences. Probably most people have at least a general intuitive idea of what a sequence is. One simple example of a sequence is

$$\{1, 2, 3, 4, 5, \ldots\},$$

while another more irregular sequence is

$$\{3, 7, \pi, 2/5, -6, \ldots\}.$$

In the second case, it is not possible to predict what the next entry will be after -6, but nonetheless, both examples seem to satisfy the essential sense of what a sequence is. What is needed now is an exact mathematical definition:

Definition. A **sequence** is a function $a : A \to \mathbb{R}$ where either $A = \mathbb{Z}^+ := \{1, 2, 3, \ldots\}$ or $A = \mathbb{N} := \{0, 1, 2, 3, \ldots\}$ (the *natural numbers*) or perhaps A is any countable set.[1] Thus, either

$$a : \{a_n\} = \{a_1, a_2, a_3, \ldots\}$$

or

$$a : \{a_n\} = \{a_0, a_1, a_2, a_3, \ldots\},$$

where in both cases $a_n \in \mathbb{R}$. We refer to a_n as the n-th **element** or **entry** of the sequence.

Remarks.
1. Notice that a sequence differs from a set in that a sequence has an order, whereas a set is simply a collection of elements. Technically, there is no first element in a set, but there is in a sequence.
2. This is not the most general definition of the word "sequence," but it is easily general enough for our purposes. It contains all the key ideas to discuss sequences. In particular, we will not discuss finite sequences.
3. Notice that for our purposes, sequences begin either with $n = 0$ or with $n = 1$. Beginning with a zeroth element is less common, but it occurs sometimes when a zeroth element makes sense. Among other places, this happens in computer science. Still, unless there is some specific reason to do otherwise, let $A = \mathbb{Z}^+$.

A sequence can be depicted graphically on the real line as in Figure 1.1.

---

1 In mathematics, *countable* does not generally mean finite. Rather it means that there is a first element, then a second element, then a third and so forth, probably going on without ending.

https://doi.org/10.1515/9783110768398-001

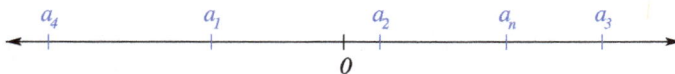

Figure 1.1: A sequence on the real line. Here several elements of the sequence are shown in blue. Notice that there is no particular pattern to this sequence. Which of the elements are positive, and which are negative?

1.1 Convergent sequences

Now that we have a definition for sequences, we can introduce the central concept of convergence, starting with a basic example.

Example 1.1.1. Suppose that $a_n := 1/n$ (i. e., a_n is *defined* equal to $1/n$), so that

$$\{a_n\} = \{1/n\} = \left\{1, \frac{1}{2}, \frac{1}{3}, \frac{1}{4}, \ldots\right\}.$$

Notice that this sequence must start with $n = 1$ since a_0 is undefinable using the given formula. Also notice that a_n decreases toward zero as n increases toward infinity, even though $a_n \neq 0$ for any $n \in \mathbb{Z}^+$. In symbols, that is $a_n \searrow 0$ as $n \nearrow \infty$.

All of this may seem clear intuitively, but as we will see, intuition is not always so clear. So again an exact mathematical definition is needed.

Definition. A sequence $\{a_n\} \subset \mathbb{R}$ **converges** to a limit $L \in \mathbb{R}$ iff given $\varepsilon > 0$, $\exists N \in \mathbb{Z}^+$ such that $|a_n - L| < \varepsilon$ whenever $n > N$. Written with a minimal amount of symbols, this definition is "a sequence $\{a_n\}$ of real numbers **converges** to a limit L in the reals if and only if given $\varepsilon > 0$, there exists N in the positive integers such that $|a_n - L| < \varepsilon$ whenever $n > N$." The sequence $\{a_n\}$ **diverges** iff it does not converge.

Notation. The notation to indicate that a sequence $\{a_n\}$ converges to a limit L is

$$\lim_{n \to \infty} a_n = L,$$

or more briefly

$$\{a_n\} \to L \quad \text{or} \quad a_n \to L.$$

Remark. In this definition, ε can be any positive real number, but it is generally thought of as a *small* positive real number. The definition works because ε can be an arbitrary small positive real number. A schematic diagram of the relationship between ε, L and the elements of the sequence a_k is shown in Figure 1.2.

Example 1.1.1. Returning to our first example, suppose again that $a_n := 1/n$. Does the sequence $\{a_n\}$ converge, and if so, to where? For a given ε, what value of N guarantees that the convergence definition is satisfied?

Figure 1.2: A sequence approaching a limit L on the real line. The early elements of the sequence through a_N can be outside the interval $(L - \varepsilon, L + \varepsilon)$, but starting with a_{N+1}, all of the elements must lie inside the interval. In particular, when $n > N$, then a_n must lie in the interval. Where might a_{N+5} be on this number line? If ε is made smaller, then likely N will have to be made larger to keep the elements of the sequence from a_{N+1} onward in the new smaller interval.

Answer. Notice that there are two questions here: The first is a calculus question (Where does the sequence converge to? What is the value of L?), while the second is an analysis question (How does one show that the definition is satisfied?). In this case, the calculus question is relatively easy, and in fact we have already mentioned the answer: $L = 0$. The second question is more interesting: Given $\varepsilon > 0$, what value of N is needed to guarantee that $|a_n - L| < \varepsilon$ when $n > N$? Notice that here

$$|a_n - L| = |a_n - 0| = |a_n| = 1/n < \varepsilon \quad \Longleftarrow \quad n > 1/\varepsilon.$$

This may seem obvious, but notice that in fact, it gives us the answer we are looking for: define $N := \lceil 1/\varepsilon \rceil$, where $\lceil x \rceil$ is the ceiling function or next greater integer function for x (the least integer greater than or equal to x). It is important that this ceiling function be used because N must be an integer, but for most values of ε, its reciprocal $1/\varepsilon$ is not itself an integer. Then for this example, $|a_n - L| < \varepsilon$ whenever $n > N := \lceil 1/\varepsilon \rceil \geq 1/\varepsilon$.

In summary, the limit here is $L = 0$ and setting $N := \lceil 1/\varepsilon \rceil$ allows the definition of convergence to be satisfied.

Remarks.
1. Notice that the value of N increases as ε decreases; this is almost always the case. Expect ε to be in the denominator in the expression for N.
2. Notice the direction of the double arrow above. Although we started on the left working on $|a_n - L|$, we need to keep in mind that this is the conclusion to be reached provided that n is large enough. So a backward implication (a backward double arrow) is what is needed. Indeed, the symbol "$\Longleftarrow$" is perhaps best read as "whenever". Thus, the implication above becomes $|a_n - L| < \varepsilon$ whenever $n > 1/\varepsilon$.

The first example above is relatively straightforward; the next two examples are more interesting ("more interesting" means "harder").

Example 1.1.2. Suppose that

$$a_n := \frac{2n^2}{3n^2 + 1}.$$

Again, (calculus question) to what value L does this sequence converge, and (analysis question) for a given $\varepsilon > 0$, how large must n be if the limit definition is to be satisfied?

Answer. First consider the calculus question. Again, we must start by finding (if we can) the value of the limit L. This time, it may not be quite so obvious since both the numerator and the denominator in a_n go to infinity as $n \to \infty$. Sequences of this form are said to be in the indeterminate form "∞/∞" (the quotation marks indicate that one does not think of this form as a normal fraction since the exact value of a limit in this form can be anything from 0 to ∞). By dividing both the numerator and the denominator by n^2 (the highest-degree term in either the numerator or denominator), however, one finds that

$$a_n = \frac{2n^2}{3n^2 + 1} = \frac{2}{3 + 1/n^2}$$

and in this new form, it is clear that $a_n \to 2/3$ as $n \to \infty$.

Now we turn to the analysis question. For a given $\varepsilon > 0$, we need

$$|a_n - L| = \left| \frac{2n^2}{3n^2 + 1} - \frac{2}{3} \right| < \varepsilon.$$

Finding a common denominator to combine these fractions, one sees that we need

$$|a_n - L| = 2 \left| \frac{3n^2 - (3n^2 + 1)}{3(3n^2 + 1)} \right| = \frac{2}{3} \left(\frac{1}{3n^2 + 1} \right) < \varepsilon.$$

From here, one can solve for n^2 to find that the above inequality is equivalent to

$$\frac{2}{9\varepsilon} - \frac{1}{3} < n^2,$$

but this leads to a small problem: The expression on the left could be negative if ε is bigger than 2/3, and this would be a problem when we take the square root of both sides to find a lower bound for n. Notice that this problem can be avoided, however, if we make the lower bound on n^2 a bit bigger:

$$\frac{2}{9\varepsilon} - \frac{1}{3} < n^2 \quad \Longleftarrow \quad \frac{2}{9\varepsilon} < n^2.$$

The above implication illustrates an important part of analysis: We often work with inequalities and approximations rather than always working with equalities. Sometimes one can simply drop inconvenient terms while preserving the implication that is needed. Now taking the square root of both sides, one can find a lower bound for n that guarantees that the inequality $|a_n - L| < \varepsilon$ that we need is satisfied:

$$\frac{\sqrt{2}}{3\sqrt{\varepsilon}} < n.$$

This lower bound on n lets us define N. To guarantee an integral value, the ceiling function can again be used:

$$N := \left\lceil \frac{\sqrt{2}}{3\sqrt{\varepsilon}} \right\rceil.$$

So in summary, turning the implications around so the arrows point forward, we have

$$n > N := \left\lceil \frac{\sqrt{2}}{3\sqrt{\varepsilon}} \right\rceil \implies \frac{2}{9\varepsilon} < n^2 \implies \frac{2}{9\varepsilon} - \frac{1}{3} < n^2$$

$$\implies |a_n - L| = \left| \frac{2n^2}{3n^2 + 1} - \frac{2}{3} \right| < \varepsilon,$$

and these inequalities guarantee that for this sequence, $\{a_n\} \to 2/3$, so the definition is satisfied.

Although what was said above really answers the question, it may be helpful in this example to look at the values of a few elements in this sequence:

$$\{a_n\} = \{a_1, a_2, a_3, a_4, a_5, a_6, \ldots, a_{10}, \ldots, a_{100}, \ldots\}$$
$$= \{1/2, 8/13, 9/14, 32/49, 25/38, 72/109, \ldots, 200/301, \ldots, 20000/30001, \ldots\}$$
$$\approx \{0.5, 0.61538, 0.64285, 0.65306, 0.65789, 0.66055, \ldots, 0.66445, \ldots 0.66664, \ldots\}.$$

This sequence is shown graphically in Figure 1.3; notice that this sequence seems to be increasing and bounded above by the limiting value 2/3.

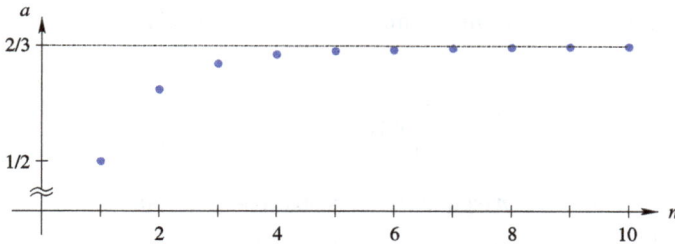

Figure 1.3: The first few elements of the sequence $\{2n^2/(3n^2 + 1)\}$; notice that it increases toward the limiting value, 2/3.

Example 1.1.3. Suppose that $a_n := n(\sqrt{n^2 + 1} - n)$. Does this sequence converge, and if so, to what value L? Given $\varepsilon > 0$, how large must n be if the limit definition is to be satisfied?

Answer. This time, neither the calculus question (What's the limit L?) nor the analysis question (How big must n be for a_n to be within ε of L?) is easy.

How should one compute/find a limit when the result is not obvious? The answer depends on the indeterminate form of the limit. Notice that this limit is of the form "$\infty \cdot (\infty - \infty)$": n and $\sqrt{n^2 + 1}$ both go to infinity, making this a complicated limit. To compute the value of L, the key is to creatively multiply by 1, that is, to multiply and divide a_n by the *conjugate* of the expression in parentheses:

$$a_n = n(\sqrt{n^2+1} - n)\frac{\sqrt{n^2+1}+n}{\sqrt{n^2+1}+n} = \frac{n(n^2+1-n^2)}{\sqrt{n^2+1}+n} = \frac{n}{\sqrt{n^2+1}+n} = \frac{1}{\sqrt{1+1/n^2}+1}.$$

As $n \to \infty$, the term $1/n^2 \to 0$. Thus, $a_n \to L = 1/2$ (bet y' didn't see that comin').

As before, once the value of L is known, one can show that the definition of convergence is satisfied by finding the value of N as a function of ε. So the question is, how large must n be to make

$$\left| n(\sqrt{n^2+1} - n) - \frac{1}{2} \right|$$

smaller than ε? Again multiplying and dividing by the conjugate, one finds that

$$|a_n - L| = \left| n(\sqrt{n^2+1} - n)\frac{\sqrt{n^2+1}+n}{\sqrt{n^2+1}+n} - \frac{1}{2} \right| < \varepsilon,$$

which is equivalent to

$$\left| \frac{1}{\sqrt{1+1/n^2}+1} - \frac{1}{2} \right| < \varepsilon,$$

or after finding a common denominator and combining the two terms, it is equivalent to

$$\left| \frac{1 - \sqrt{1+1/n^2}}{1 + \sqrt{1+1/n^2}} \right| < 2\varepsilon.$$

In this case, it is impossible to solve for n exactly, but it is possible to simplify and adjust this inequality to find a bound on n that gives us what we need. Multiplying the numerator and denominator of the previous inequality by the conjugate of the numerator and then simplifying, one obtains another equivalent form of the inequality:

$$\frac{1/n^2}{(1 + \sqrt{1+1/n^2})^2} < 2\varepsilon.$$

From this final equivalent inequality, one still cannot solve for n, but one can adjust the bound to find an inequality that can be solved for n. The key is making the left hand side of the previous inequality slightly *larger* by dropping the $1/n^2$ term in the denominator (notice that making the denominator smaller makes the entire left hand side larger). Now notice that if one bounds this larger expression by 2ε, one gets the desired bound for free:

$$\frac{1/n^2}{(1 + \sqrt{1 + 1/n^2})^2} < \frac{1/n^2}{(1 + \sqrt{1 + 0})^2} = \frac{1}{4n^2} < 2\varepsilon.$$

Finally, the bound on n we want is

$$\frac{1}{8\varepsilon} < n^2 \iff \frac{1}{2\sqrt{2\varepsilon}} < n,$$

so N can be defined as

$$N := \left\lceil \frac{1}{2\sqrt{2\varepsilon}} \right\rceil.$$

1.2 Computing sequence limits

The previous section defined the limit of a sequence and discussed how to use that definition to firmly establish that various sequences really do converge to the limits that they seem to converge to. This can be thought of as the analysis question involving limits and sequences. Now we turn our attention to what can be thought of as the calculus question: just finding L, the value of the limit, without using our definition to prove that the limit really does exist. As mentioned above, this is actually the first part of each of the three examples discussed in the previous section. Sometimes this question is relatively easy, as it was for the sequence where $a_n = 1/n$. The next several examples deal with cases where the calculus question is more than trivial to answer. Several additional indeterminate forms are introduced.

Example 1.2.1. Please find

$$\lim_{n \to \infty} \frac{5n^3 - 2n}{n^2 + 7}$$

provided that this limit exists.

Answer. Notice that this question is similar to the one in Example 1.1.2 (it is again of the form "∞/∞"), but of course here the powers of n in the rational expression are different. Still, the approach used in Example 1.1.2 works here too: We divide by the highest power of n in the rational expression. We obtain

$$\lim_{n \to \infty} \frac{5n^3 - 2n}{n^2 + 7} = \lim_{n \to \infty} \frac{5 - 2/n^2}{1/n + 7/n^3}.$$

Then, taking $n \to \infty$, one sees that the three terms with reciprocal powers of n all go to 0, meaning that the numerator goes to 5 while the denominator goes to 0 through positive numbers; hence,

$$\frac{5n^3 - 2n}{n^2 + 7} \to +\infty.$$

This is a form of divergence (not convergence since L must be a real number) called *divergence to infinity*; this term will be defined mathematically below.

Example 1.2.2. Next consider the sequence whose n-th entry is

$$a_n := e^{-n} \sin n = \frac{\sin n}{e^n}.$$

Please find its limit, provided that this limit exists.

Answer. This expression may look more difficult to evaluate because of the trigonometric and exponential functions, and indeed one cannot just compute the limit of the numerator and the limit of the denominator separately, in part because the limit of the numerator does not exist. Still, this is *not* an indeterminate form, and in fact this limit is relatively easy to find. First notice that $\forall\, n \in \mathbb{Z}^+$,

$$-1 \le \sin n \le 1,$$

so since $e^{-n} > 0$,

$$-e^{-n} \le e^{-n} \sin n \le e^{-n}.$$

Since $e^{-n} \to 0$ as $n \to \infty$, the expression at hand must also converge to 0 as $n \to \infty$.

The above example demonstrates the use of the sandwich theorem, also known as the squeeze theorem, which is discussed and proven below.

Example 1.2.3. Find the value that the sequence

$$a_n := \frac{\ln n}{n},$$

approaches, provided that this limit exists.

Answer. The reader is assumed to know a basic property of the logarithm function ln: that it grows without bound as $n \to \infty$. Thus, the expression $\ln n/n$ is again of the form "∞/∞" as $n \to \infty$. Again, one might expect that this is a difficult limit to evaluate, but there is a standard approach that is often helpful for limits of the indeterminate forms "$0/0$" or "∞/∞": l'Hôpital's rule. According to the l'Hôpital[2] rule, one can evaluate limits of either of these two forms by differentiating the numerator and the denominator separately and then evaluating that limit. Symbolically, this is

$$\lim_{n \to \infty} \frac{f(n)}{g(n)} = \lim_{n \to \infty} \frac{f'(n)}{g'(n)},$$

---

2 Guillaume de l'Hôpital (1661–1704) was a French mathematician who in 1696 published what was arguably the first differential calculus textbook; it contained the rule that bears his name.

where the derivatives are taken with respect to n. Normally l'Hôpital's rule is introduced and (one hopes) proven in basic calculus courses, so it is taken for granted here. In this particular example, l'Hôpital's rule implies that

$$\lim_{n\to\infty} \frac{\ln n}{n} = \lim_{n\to\infty} \frac{1/n}{1} = \lim_{n\to\infty} \frac{1}{n} = 0.$$

1.2.1 Two famous limits

There are a couple of famous limits whose values should be known before one considers the rest of the examples in this section.

Theorem 1.2.1.

$$\lim_{\theta\to 0} \frac{\sin\theta}{\theta} = 1,$$

$$\lim_{n\to\infty} \left(1 + \frac{1}{n}\right)^n = e.$$

Proof. One might expect that the proof of the first would be a simple application of the l'Hôpital rule, but such an approach would be circular since the value of this limit must be known to differentiate the sine function. So the proof of the first, which is rather geometric, should be given in any basic calculus text.

As for the second, what specifically needs to be proven depends on how one sets up certain definitions. Here let us define the natural logarithm function to be the following famous and important integral:

$$\ln x := \int_1^x \frac{dt}{t}, \quad x > 0.$$

Section 5.1 below has more on $\ln x$; for now, suffice it to say that the number e is the base of the natural logarithm ($\ln x \equiv \log_e x$), and the exponential function $\exp(x) \equiv e^x$ is the inverse of the natural logarithm function, which must exist since $\ln x$ is strictly increasing. One can then show that the exponential function is (up to a constant multiple) the only function whose derivative is itself. Using this integral definition of the logarithm function, one can see that this function must be continuous, and therefore by applying l'Hôpital's rule, we have

$$\ln\left(\lim_{n\to\infty}\left(1 + \frac{1}{n}\right)^n\right) = \lim_{n\to\infty} \ln\left(1 + \frac{1}{n}\right)^n = \lim_{n\to\infty} n\ln\left(1 + \frac{1}{n}\right)$$

$$= \lim_{n\to\infty} \frac{\ln(1 + \frac{1}{n})}{\frac{1}{n}} = \lim_{n\to\infty} \frac{\frac{1}{(1+\frac{1}{n})}(-\frac{1}{n^2})}{(-\frac{1}{n^2})} = \lim_{n\to\infty} \frac{1}{(1 + \frac{1}{n})} = 1.$$

Notice that it is the continuity of the logarithm function that implies that the logarithm of the limit is the limit of the logarithm. Since $\ln u = 1$ implies that $u = e$, the result is proven. $\square$

The technique used in the previous proof as well as the results of the above theorem can be used to solve the next examples.

Example 1.2.4. Please find the limit of the sequence

$$a_n := n \sin(1/n)$$

if the limit exists.

Answer. As written, this limit is of the form "$\infty \cdot 0$." Notice that if $\theta = 1/n$, then by substitution

$$\lim_{n \to \infty} n \sin(1/n) = \lim_{n \to \infty} \frac{\sin(1/n)}{1/n} = \lim_{\theta \to 0} \frac{\sin \theta}{\theta} = 1$$

because of the first part of the previous theorem. One can also confirm this result by using l'Hôpital's rule after putting the limit into "0/0" form.

Example 1.2.5. Evaluate

$$\lim_{n \to \infty} \left(\frac{n+5}{n+1} \right)^n$$

provided this limit exists.

Answer. Notice that this limit, like the second one in Theorem 1.2.1, is of the form "1^∞." For limits of this form, as well as those of the forms "∞^0" and "0^0," the technique used in the previous proof or the theorem itself can be used to resolve the value of the limit:

$$
\begin{aligned}
\ln\left(\lim_{n \to \infty} \left(\frac{n+5}{n+1} \right)^n \right) &= \lim_{n \to \infty} n \ln\left(\frac{n+5}{n+1} \right) = \lim_{n \to \infty} n \ln\left(\frac{1+5/n}{1+1/n} \right) \\
&= \lim_{n \to \infty} n\left(\ln(1+5/n) - \ln(1+1/n) \right) \\
&= \lim_{n \to \infty} n \ln(1+5/n) - \lim_{n \to \infty} n \ln(1+1/n) \\
&= \lim_{n \to \infty} \ln\left((1+5/n)^n \right) - \lim_{n \to \infty} \ln(1+1/n)^n \\
&= \ln\left(\lim_{m \to \infty} (1+1/m)^{5m} \right) - \ln\left(\lim_{n \to \infty} (1+1/n)^n \right) \\
&= \ln\left(\left(\lim_{m \to \infty} (1+1/m)^m \right)^5 \right) - \ln e = 5 \ln e - 1 = 4.
\end{aligned}
$$

So the limit in question must approach e^4. Notice that in the first of the two limits, the substitution $n = 5m$ is used.

To conclude this section, it is perhaps helpful to simply list the standard indeterminate forms: (1) "0/0," (2) "∞/∞," (3) "$0 \cdot \infty$," (4) "1^∞," (5) "∞^0," (6) "0^0," and (7) "$\infty - \infty$." The first three of these may be resolved using the l'Hôpital rule, while the next three likely require the introduction of the logarithm, and the last likely requires conjugate multiplication.

1.3 Divergence to infinity

So far, we have focused on convergence and discussed divergence only as the opposite to convergence. There is one type of divergence, however, that deserves special attention: divergence to infinity (positive or negative infinity). In this case, the elements of the sequence eventually become either unboundedly large or unboundedly negative. The following definition makes this idea exact.

Definition. A sequence of real numbers $\{a_n\}$ **diverges to infinity** ($a_n \to +\infty$) iff given $M > 0$, $\exists N \in \mathbb{Z}^+$ such that $a_n > M$ whenever $n > N$. The sequence $\{a_n\}$ **diverges to negative infinity** ($a_n \to -\infty$) iff the sequence $\{-a_n\}$ diverges to infinity.

Remark. In the definition of convergence, ε is used because it is one of the smallest letters in the Greek alphabet, and we think of ε as having a very small value—much less than 1. Here M is used because it is one of the largest letters in the Latin alphabet, and we think of M as having a very large value—much greater than 1.

Although this definition is completely self-contained and exact, it helps if there are examples to illustrate it. One such example was Example 1.2.1 above. Another (this one requiring us to find N in terms of M) follows next.

Example 1.3.1. Suppose that

$$a_n := \frac{2n^2}{3n+1}.$$

Please show that this sequence $\{a_n\}$ diverges to infinity as $n \to \infty$.

Answer. This example looks very similar to Example 1.2.1, but now we are asked to show that the definition of diverges to infinity is satisfied. Here the numerator is quadratic, while the denominator is affine-linear. This means that as n gets large, even though the ratio is of the form "∞/∞," the numerator is growing more quickly. Indeed, this outcome can also be seen by dividing the numerator and denominator by n:

$$a_n := \frac{2n^2}{3n+1} = \frac{2n}{3+1/n}.$$

Now as n gets large, the numerator also gets large, but the denominator approaches 3, so the ratio (the rational expression) on the whole becomes large.

Now consider the definition: Given $M > 0$, how large must n be if a_n is always greater than M? Notice that

$$a_n := \frac{2n^2}{3n+1} = \frac{2n}{3+1/n} > M \quad \Longleftarrow \quad \frac{2n}{3+1/n} \geq \frac{2n}{3+1} > M$$

since $1/n \leq 1 \,\forall\, n \in \mathbb{Z}^+$ (making the denominator a bit bigger makes the rational expression smaller and allows us to find a lower bound on n). But of course,

$$\frac{2n}{3+1} > M \quad \Longleftrightarrow \quad n > M/2,$$

which means that a good choice for N is $N := \lceil M/2 \rceil$. Then if $n > N$, it follows that $a_n > M$.

1.4 Convergence implications: sequence convergence theorems

The analysis presented above might make one think that there is a strict definition primarily as a check on our computations or guesses of the possible values of L. This view is at best only partially true. The more important reason for having strict, mathematical definitions is to establish general results/theorems—statements that can be proven to always be true. The first theorem below shows that any sequence converges to at most one limit L, so if an L is known to exist, there is no need to look for another. The second allows us to use convergence information about certain sequences and transfer it to other sequences, while the third gives us a test that can help us to determine whether or not a sequence converges.

1.4.1 Limit uniqueness

Theorem 1.4.1 (Uniqueness of L). *For any sequence $\{a_n\}$, if the limit L exists, then it is unique.*

Remark. This is a uniqueness theorem (it claims that there is at most one of something). The proofs of uniqueness theorems generally follow a certain pattern. One assumes that there are two, and shows that the two are in fact the same.

Proof. Suppose that $\{a_n\}$ converges to a limit L_1 and also converges to a limit L_2. Given $\varepsilon > 0$, since $\{a_n\}$ converges to L_1, by the definition of limit, $\exists\, N_1 \in \mathbb{Z}^+$ such that

$$|a_n - L_1| < \frac{\varepsilon}{2}$$

$\forall\, n > N_1$. But also $\{a_n\}$ converges to L_2, so $\exists\, N_2 \in \mathbb{Z}^+$ such that

$$|a_n - L_2| < \frac{\varepsilon}{2}$$

$\forall\, n > N_2$. Let $N := \max\{N_1, N_2\}$. Then by the triangle inequality, $\forall\, n > N$,

$$|L_1 - L_2| = \left|(L_1 - a_n) + (a_n - L_2)\right| \le |a_n - L_1| + |a_n - L_2| < \frac{\varepsilon}{2} + \frac{\varepsilon}{2} = \varepsilon.$$

Since ε can be any *arbitrarily* small positive real numbers, this inequality can only be satisfied if $|L_1 - L_2| = 0 \implies L_1 = L_2$. □

Remark. This proof uses the triangle inequality, which states that for any three real numbers a, b and c,

$$|a - b| \le |a - c| + |c - b|.$$

The name comes from geometry, where the length of any side of a triangle is less than the sum of the lengths of the other two sides. The proof of the version of the triangle inequality used here is given in Exercise B.1.3.

The above uniqueness theorem has a number of important implications. One of these is that if two *subsequences* (this term will be defined mathematically below) inside a given sequence are heading for two separate limits, then the sequence itself cannot converge. The next example will demonstrate this result.

Example 1.4.1. Consider the sequence $\{a_n\}$, where

$$a_n = (-1)^n + 1/n.$$

Does this sequence converge? If so, to what limit L?

Answer. Notice that this sequence is really comprised of two distinct subsequences: The even elements are $a_{2n} = 1 + 1/2n$ and the odd elements are $a_{2n+1} = -1 + 1/(2n + 1)$. By inspection, the subsequence of even elements converges to 1, while the subsequence of odd elements converges to −1. Since part of the sequence is approaching $L_1 = 1$, but another part is approaching $L_2 = -1$, there cannot be a single unique limit for the entire sequence, and hence in this case, the limit does not exist. So this sequence can be thought of as an example of the uniqueness theorem in action: If different parts of a sequence converge to separate values, this implies that the limit does not exist.

1.4.2 Sandwich theorem

The next theorem considers a sequence sandwiched between two other sequences that both converge to the same limit; it guarantees that the sequence in the middle must also converge to that limit.

Theorem 1.4.2 (Sandwich theorem). *Given three sequences of real numbers $\{a_n\}$, $\{b_n\}$ and $\{c_n\}$, suppose that $a_n \le b_n \le c_n \ \forall n \in \mathbb{Z}^+$ and that $\{a_n\} \to L$ and $\{c_n\} \to L$ for some $L \in \mathbb{R}$. Then $\{b_n\} \to L$ (the sequence $\{b_n\}$ must also converge to L).*

Proof. Given $\varepsilon > 0$, since $\{a_n\}$ converges to L, there exists $N_a \in \mathbb{Z}^+$ such that $|a_n - L| < \varepsilon$ whenever $n > N_a$. But also, since $\{c_n\}$ converges to L, there exists $N_c \in \mathbb{Z}^+$ such that $|c_n - L| < \varepsilon$ whenever $n > N_c$. Define $N := \max\{N_a, N_c\}$. Now $|a_n - L| < \varepsilon$ is equivalent to $|L - a_n| < \varepsilon$ and therefore implies that $L - \varepsilon < a_n$. At the same time $|c_n - L| < \varepsilon$ implies that $c_n < L + \varepsilon$. Recalling that $a_n \le b_n \le c_n$, one finds that

$$L - \varepsilon < a_n \le b_n \le c_n < L + \varepsilon,$$

and thus on subtracting L across this inequality, one finds that $-\varepsilon < b_n - L < \varepsilon$. This inequality then is equivalent to $|b_n - L| < \varepsilon$, and since this final inequality holds $\forall n > N$, by definition, the sequence $\{b_n\}$ converges to L. $\qquad\square$

An example of how this theorem can be used has already been discussed; see Example 1.2.2 above. There is also a version of the sandwich theorem that applies to divergence to infinity:

Theorem 1.4.3. *Given two sequences of real numbers $\{a_n\}$ and $\{b_n\}$, suppose that $a_n \le b_n \ \forall n \in \mathbb{Z}^+$ and that $\{a_n\} \to \infty$. Then $\{b_n\} \to \infty$. (The sequence $\{b_n\}$ must also diverge to infinity if the sequence $\{a_n\}$ does.)*

The proof of the previous theorem is left as an exercise—see Exercise B.1.4.

1.4.3 Sequence ratio test

It is perfectly acceptable to show that a sequence converges by appealing directly to the definitions as was done above, but there are also times when it is useful to have more general theorems that allow us to answer convergence questions without having to go back to the definition. One such theorem is the ratio test for sequences (there will be another version of the ratio test later for series).

Perhaps the best way to justify this ratio test is to consider an example where the limit would be difficult to compute using the techniques that have been introduced so far.

Example 1.4.2. Find the limit value L for the sequence whose n-th element is

$$a_n = \frac{2^n}{n!}.$$

Answer. Notice that this limit has a factorial in its denominator, and factorials are generally only defined for integers, not for real numbers, i. e., one usually does not write $x!$

when x is a real number that is not an integer. Because of this, there really is no way to apply the l'Hôpital rule to this limit.[3] Also, there is no helpful conjugate expression here. It is true that one can find the value of this limit by factoring both the numerator and the denominator, but there is perhaps an even easier approach using the ratio test.

Theorem 1.4.4 (Sequence ratio test). *If $\{a_n\}$ is any sequence with $a_n \neq 0 \ \forall n \in \mathbb{Z}^+$ and $\lim_{n\to\infty} \left|\frac{a_{n+1}}{a_n}\right| = r \in [0, +\infty)$, then:*
(a) *If $r < 1$, then $a_n \to 0$.*
(b) *If $r > 1$, then $|a_n| \to \infty$.*
(c) *If $r = 1$, then the ratio test fails (the sequence may converge or diverge).*

Answer. Returning to our previous example and its answer, one can apply the sequence ratio test by considering the limit of the ratio:

$$\lim_{n\to\infty}\left|\frac{a_{n+1}}{a_n}\right| = \lim_{n\to\infty}\frac{a_{n+1}}{a_n} = \lim_{n\to\infty}\frac{\frac{2^{n+1}}{(n+1)!}}{\frac{2^n}{n!}} = \lim_{n\to\infty}\frac{2^{n+1}n!}{2^n(n+1)!} = \lim_{n\to\infty}\frac{2}{n+1} = 0.$$

Thus, by the ratio test, since the limit of the ratio exists and is less than 1, the original sequence converges to 0:

$$\lim_{n\to\infty} a_n = \lim_{n\to\infty}\frac{2^n}{n!} = 0.$$

Every good theorem requires a proof, and the sequence ratio test is no exception.

Proof. (a) Suppose first that $r < 1$. The assumptions suggest that we need to get n large enough so that the ratio is close to r, but how should we take advantage of this? Suppose that $\rho := (r + 1)/2$ (so ρ is the midpoint between 1 and r). Consider the diagram in Figure 1.4. Notice that since the limit of the ratio converges to r, there exists $N \in \mathbb{Z}^+$ such that

$$\left|\frac{a_n}{a_{n-1}}\right| < \rho$$

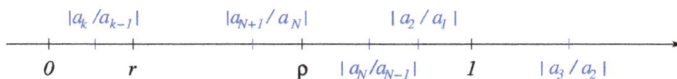

Figure 1.4: Sequence ratios converging to r. There is no particular control on the ratios for the first few elements in the sequence, but for $n > N$ ($n = N+1$ and $n = k$ in this diagram), each ratio is less than ρ. What is ε in this case, and where should it be placed on this diagram?

---

3 It is true that factorials can be expressed as gamma functions, and these are generally differentiable, but it is difficult to deal with gamma functions and their derivatives, so they really do not make l'Hôpital's rule useful here.

whenever $n > N$. So the ratio for $n = N + 1$ is the first element of the sequence of ratios that must be less than ρ, and this must also be true for every ratio for $n > N + 1$.

But now that we have our ratio smaller than ρ whenever n is sufficiently large, how do we use this to complete our proof? One of the most important things that a mathematician can do in some situations is to either add zero or multiply by one, but to do it in a creative, helpful way. In this case, notice that since $a_n \neq 0 \; \forall n \in \mathbb{Z}^+$, when $n > N$,

$$
\begin{aligned}
|a_n| &= \left| \frac{a_n}{a_{n-1}} \cdot \frac{a_{n-1}}{a_{n-2}} \cdot \frac{a_{n-2}}{a_{n-3}} \cdots \frac{a_{N+2}}{a_{N+1}} \cdot \frac{a_{N+1}}{a_N} \cdot a_N \right| \\
&= \left| \frac{a_n}{a_{n-1}} \right| \left| \frac{a_{n-1}}{a_{n-2}} \right| \left| \frac{a_{n-2}}{a_{n-3}} \right| \cdots \left| \frac{a_{N+2}}{a_{N+1}} \right| \left| \frac{a_{N+1}}{a_N} \right| |a_N| .
\end{aligned}
$$

We are using the fact that the absolute value of a product equals the product of the absolute values: $|xy| = |x||y|$. Now each of the ratios is bounded by ρ, so

$$
\begin{aligned}
|a_n| &= \left| \frac{a_n}{a_{n-1}} \cdot \frac{a_{n-1}}{a_{n-2}} \cdot \frac{a_{n-2}}{a_{n-3}} \cdots \frac{a_{N+2}}{a_{N+1}} \cdot \frac{a_{N+1}}{a_N} \cdot a_N \right| \\
&= \left| \frac{a_n}{a_{n-1}} \right| \left| \frac{a_{n-1}}{a_{n-2}} \right| \left| \frac{a_{n-2}}{a_{n-3}} \right| \cdots \left| \frac{a_{N+2}}{a_{N+1}} \right| \left| \frac{a_{N+1}}{a_N} \right| |a_N| \\
&< \rho \cdot \rho \cdot \rho \cdots \rho \cdot \rho \cdot |a_N| \\
&= \rho^{n-N} |a_N| = \rho^n \frac{|a_N|}{\rho^N} .
\end{aligned}
$$

The notation $\cdots$ is needed here, but it leaves open the question of how many factors of ρ there are. How does one see that there are $n - N$ factors of ρ? Maybe the easiest way to count them in this case is to notice that the indices in the denominators start on the left with $n - 1$, followed by $n - 2$, and so forth, all the way to $N = n - (n - N)$. So we can use these indices to see that there are $n - N$ factors.

This last bound on $|a_n|$ is useful because N and a_N are both fixed for a given value of ρ, but n can be as large as we want. So since $\rho < 1$, this bound guarantees that $|a_n|$ goes to zero, and since an absolute value is zero only when its argument is zero, a_n also goes to zero.

The proof of (b) is Exercise B.1.5(a); the proof of (c) is by examples and is Exercise B.1.5(b). □

The ease with which the ratio test deals with the previous example might lead one to think that it can always or almost always work; unfortunately this is not the case, as the following simple example shows.

Example 1.4.3. Does the limit

$$
\lim_{n \to \infty} \frac{7n^2 - n + 2}{3n^2 + 9n - 7}
$$

converge, and if so, to what value?

Answer. Suppose one tries to use the sequence ratio test on the rational expression in this limit by computing the appropriate ratio test limit:

$$\lim_{n\to\infty}\frac{a_{n+1}}{a_n} = \lim_{n\to\infty}\frac{\frac{7(n+1)^2-(n+1)+2}{3(n+1)^2+9(n+1)-7}}{\frac{7n^2-n+2}{3n^2+9n-7}} = \lim_{n\to\infty}\frac{(7(n+1)^2-(n+1)+2)(3n^2+9n-7)}{(7n^2-n+2)(3(n+1)^2+9(n+1)-7)}$$

$$= \lim_{n\to\infty}\frac{(7(1+1/n)^2-(1/n+1/n^2)+2/n^2)(3+9/n-7/n^2)}{(7-1/n+2/n^2)(3(1+1/n)^2+9(1/n+1/n^2)-7/n^2)} = \lim_{n\to\infty}\frac{21}{21} = 1.$$

This of course means that the ratio test is inclusive in this case. There was no mistake in our calculation; the mistake was in our choice of methods. Indeed, the ratio test is never a useful choice when considering the limit of a rational expression. Instead, of course, a much better approach is to either apply l'Hôpital's rule or divide by the highest power present in the expression:

$$\lim_{n\to\infty} a_n = \lim_{n\to\infty}\frac{7n^2-n+2}{3n^2+9n-7} = \lim_{n\to\infty}\frac{7-1/n+2/n^2}{3+9/n-7/n^2} = \frac{7}{3}.$$

Remark. It is hard to make general statements about when to use l'Hôpital's rule, when to use the ratio test, or when to try some other approach. That said, the ratio test is often the perfect tool when a_n contains exponentials and/or factorials and much less useful if there are no exponentials or factorials present and the limit involves only polynomials and rational expressions. In the end, however, the only way to know for sure whether or not a certain approach will work for a certain sequence is to try to apply it.

1.5 Limit operations

This section presents a set of results that let us find the limits of certain sequences based on arithmetic operations on other sequences and their limits.

Theorem 1.5.1 (Limit operations theorem). *Suppose that $\{a_n\}$ and $\{b_n\}$ are sequences of real numbers. If a sequence $\{a_n\}$ converges to a limit A and a sequence $\{b_n\}$ converges to a limit B (that is, $\{a_n\} \to A$ and $\{b_n\} \to B$), then:*
1. $\lim_{n\to\infty}(a_n \pm b_n) = A \pm B$.
2. $\lim_{n\to\infty}(a_n b_n) = AB$.
3. $\lim_{n\to\infty}\frac{a_n}{b_n} = \frac{A}{B}$ *provided that $B \neq 0$.*
4. *If $a_n \leq b_n \ \forall n \in \mathbb{Z}^+$, then $A \leq B$.*

Proof. (1) Given $\epsilon > 0$, since $\{a_n\} \to A$, $\exists N_A \in \mathbb{Z}^+$ such that $|a_n - A| < \epsilon/2$ whenever $n > N_A$. Since $\{b_n\} \to B$, $\exists N_B \in \mathbb{Z}^+$ such that $|b_n - B| < \epsilon/2$ whenever $n > N_B$. Define $N := \max\{N_A, N_B\}$. Then for $n > N$,

$$|(a_n + b_n) - (A + B)| = |(a_n - A) + (b_n - B)| \leq |a_n - A| + |b_n - B| < \epsilon/2 + \epsilon/2 = \epsilon.$$

Thus, $\{a_n + b_n\} \to A + B$. The same sort of proof can be used to show that the limit of the difference of two converging sequences is the difference of the limits of those sequences; see Exercise B.1.6.

(2) Given $\epsilon > 0$, since $\{a_n\} \to A$, $\exists N_A \in \mathbb{Z}^+$ such that $|a_n - A| < \frac{\epsilon}{2(1+|B|)}$ whenever $n > N_A$. Since $\{b_n\} \to B$, $\exists N_B \in \mathbb{Z}^+$ such that $|b_n - B| < \min\{\frac{\epsilon}{2(1+|A|)}, 1\}$ whenever $n > N_B$. Define $N := \max\{N_A, N_B\}$. Now add and subtract the same thing in a creative way. Then for $n > N$,

$$
\begin{aligned}
|a_n b_n - AB| &= |a_n b_n - Ab_n + Ab_n - AB| \\
&\leq |a_n b_n - Ab_n| + |Ab_n - AB| && \text{by the } \triangle\text{-inequality} \\
&= |a_n - A||b_n| + |A||b_n - B| && \text{factoring} \\
&\leq |a_n - A|(1 + |B|) + (1 + |A|)|b_n - B| && \text{creatively making the RHS bigger} \\
&< \frac{\epsilon}{2(1+|B|)}(1+|B|) + (1+|A|)\frac{\epsilon}{2(1+|A|)} && \text{insert bounds and cancel} \\
&= \frac{\epsilon}{2} + \frac{\epsilon}{2} = \epsilon.
\end{aligned}
$$

Thus, $\{a_n b_n\} \to AB$.

(3) Part (3) follows from (2) with B replaced by $1/B$ (recall here $B \neq 0$) and b_n replaced by $1/b_n$; see Exercise B.1.7.

(4) Part (4) follows from (1) by writing $a_n \neq b_n$ as $b_n - a_n \neq 0$; see Exercise B.1.8. $\square$

Remark. One might think that there could be simpler denominators in the proof for limit products than $(1 + |A|)$ and $(1 + |B|)$. But notice that one can not just divide by $|A|$ since A could be zero, and $|b_n|$ could be greater or less than $|B|$. So one must use something like $(1 + |B|)$ as the bound.

1.6 Subsequences

What if a sequence is contained in another sequence? For example, consider the sequence

$$
\{a_n\} = \left\{\frac{1}{n}\right\} = \left\{1, \frac{1}{2}, \frac{1}{3}, \frac{1}{4}, \dots\right\}
$$

and the "smaller" sequence inside it:[4]

---

4 As discussed in the next chapter, these two sequences in fact have the same number of elements; they are both *countably infinite*. Here the word "smaller" simply implies that the elements of $\{b_n\}$ form a proper subset of the elements of $\{a_n\}$.

$$\{b_n\} = \left\{\frac{1}{2n}\right\} = \left\{\frac{1}{2}, \frac{1}{4}, \frac{1}{6}, \frac{1}{8}, \ldots\right\}.$$

The latter sequence is made up of the elements of the original sequence, but only those where the denominator is even. Such a sequence $\{b_n\}$ is called a *subsequence* of the original sequence $\{a_n\}$.

There are a number of reasons to be interested in subsequences. For example, can one understand some aspects of a sequence by studying a smaller subsequence contained within? Perhaps surprisingly, the answer is yes, and this situation is considered next. But first, we need an exact definition of what a subsequence is.

Definition. Given a sequence

$$a : \{a_n\} \equiv \{a_1, a_2, a_3, \ldots\},$$

a **subsequence** is a sequence

$$b : \{b_k\} \equiv \{b_1, b_2, b_3, \ldots\}$$

where

$$b_k = a_{n_k}$$

with $k \le n_k < n_{k+1}$ for each $k \in \mathbb{Z}^+$. A **proper subsequence** is a subsequence that is not identical to the larger sequence, that is, $k < n_k$ for some $k \in \mathbb{Z}^+$.

At first glance, this definition may seem to have taken a simple concept and made it seem difficult. But sometimes one has to carefully consider a definition to understand why it hits on the key detail that allows us to be exact. In this case, requiring that $b_k = a_{n_k}$ for each $k \in \mathbb{Z}^+$ guarantees that the elements of the subsequence must come from the elements of the original sequence. The second condition ($k \le n_k < n_{k+1}$) guarantees that the subsequence may omit some elements of the sequence, but once an element is omitted, the subsequence does not return to it. In other words, each element of the subsequence comes from further down the sequence than did the previous one. Let us consider another example.

Example 1.6.1. Consider the sequence whose n-th entry is

$$a_n = \sin\left(\frac{n\pi}{4} + \frac{\pi}{n^2}\right).$$

Notice that this sequence does not converge. Please give a subsequence of this sequence that does converge.

Answer. The key to solving this example is to notice that $\pi/n^2 \to 0$ as $n \to \infty$, while the $n\pi/4$ term causes the sine function to oscillate returning to near the same value each time n increases by 8. In fact, because of the trigonometric identity

$$\sin\left(\frac{n\pi}{4} + \frac{\pi}{n^2}\right) = \sin\left(\frac{n\pi}{4}\right)\cos\left(\frac{\pi}{n^2}\right) + \cos\left(\frac{n\pi}{4}\right)\sin\left(\frac{\pi}{n^2}\right),$$

if $n = 8k$, the subsequence $b_k = a_{8k} = \sin(2\pi k)\cos(\pi/64k^2) + \cos(2\pi k)\sin(\pi/64k^2) = \sin(\pi/64k^2)$ converges to zero as k (and n) goes to $+\infty$. There are in fact many other possible converging subsequences.

The example above shows that there may be a convergent subsequence even if the original sequence diverges; the following theorem proves that if a sequence has a divergent subsequence, then the sequence itself must diverge.

Theorem 1.6.1. *If any subsequence of a given sequence diverges, then the sequence itself also diverges.*

Proof. Suppose that the given sequence is $\{a_n\}$ and that this sequence has a divergent subsequence $\{b_k\}$, where $b_k = a_{n_k}$. Since $\{b_k\}$ is divergent, given any $\epsilon > 0$, $\forall K \in \mathbb{Z}^+$, $\exists k > K$ such that $|b_k - L| > \epsilon$ no matter which $L \in \mathbb{R}$ is chosen. But $b_k = a_{n_k}$, so the same negation of the definition of convergence that was just given for $\{b_k\}$ is also possible for $\{a_n\}$, which of course implies that the given sequence diverges. □

Remark. Notice that if the subsequence diverges to infinity, then the original sequence diverges, but not necessarily to infinity.

1.7 Recognizing sequences

Until now, when we discussed specific sequences, we generally defined the sequence by giving the n-th entry in the sequence, a_n. Now we reverse this: Given the sequence $\{a_1, a_2, a_3, \ldots\}$, is it possible to write down an expression for the n-th entry, a_n? In some cases, the answer is relatively easy; for example, for the sequence $\{1, 1/2, 1/3, 1/4, 1/5, \ldots\}$, most people would say that the n-th entry is $a_n = 1/n$. In other cases, however, there is no pattern; for example, if $\{1, \pi, 47, -8, 0, 1/7, 31, \ldots\}$, it does not seem possible to give a general formula for a_n.

Even when a pattern exists, if it is not obvious, there is no general method for finding a_n. There are, however, some things to look out for.

Example 1.7.1. For the sequence $\{1/2, 1/6, 1/18, 1/54, \ldots\}$, what is the n-th entry, a_n?

Answer. Perhaps the answer in this case is also obvious to some, but if not, one can notice that each element is 1/3 of the previous element. Since $a_1 = 1/2$, the n-th entry is $a_n = 1/(2 \cdot 3^{n-1})$.

Example 1.7.2. For the sequence $\{2/9, 4/27, 8/81, 16/243, \ldots\}$, what is a_n?

Answer. The key to recognizing a_n in this case is to notice that each of the numerators is a power of 2 while the denominators are each a power of 3: $a_n = 2^n/3^{n+1}$.

Example 1.7.3. For the sequence $\{2/3, 5/9, 8/19, 11/33, \ldots\}$, what is a_n?

Answer. The n-th entry in this case is likely more difficult to spot. First notice that the numerators of consecutive entries differ by 3. The denominators, however, grow much more rapidly; this suggests that the denominator of a_n may involve some power of n. Of course, there is no guarantee, but a little experimentation yields that $a_n = (2 + 3(n-1))/(2n^2 + 1) = (3n-1)/(2n^2 + 1)$ will match at least these first four entries. Of course, if later entries do not follow this pattern, an expression for a_n might be more complicated or even impossible to find. Indeed, this might be true for any of the examples in this section.

1.8 Cauchy sequences, Cauchy convergence

The definition for convergence that was given in Section 1.1 (which is the standard definition) is extremely helpful in deriving all of the results that have been presented so far, but it does have one disadvantage: It requires us to know or determine the value of the limit L in order to check whether or not a certain sequence converges. Can a sequence be shown to converge even if it is not possible to determine L? The answer, in fact, is yes.

Definition. Suppose that $a_n \in \mathbb{R} \,\forall\, n \in \mathbb{Z}^+$. A sequence $\{a_n\}$ is a **Cauchy**[5] **sequence** if and only if given any $\epsilon > 0$, $\exists N \in \mathbb{Z}^+$ such that $|a_n - a_m| < \epsilon$ whenever $n, m > N$. A sequence satisfying this criterion (a Cauchy sequence) is said to be **Cauchy convergent**.

Example 1.8.1. Please show that the sequence $\{1/n\}$ is Cauchy convergent.

Answer. Without loss of generality (WOLOG), suppose that $m \geq n$. Then

$$|a_n - a_m| = \left|\frac{1}{n} - \frac{1}{m}\right| = \left|\frac{m-n}{mn}\right| = \frac{1 - n/m}{n} < \frac{1}{n} < \epsilon$$

whenever $n > N := \lceil \frac{1}{\epsilon} \rceil$. So this sequence is Cauchy convergent, or is a Cauchy sequence.

Of course, the sequence $\{1/n\}$ clearly converges to 0. As it turns out, *every* convergent sequence is also Cauchy convergent.

Theorem 1.8.1. *Suppose that $\{a_n\}$ is a convergent sequence. Then $\{a_n\}$ is a Cauchy sequence.*

Proof. Since $\{a_n\}$ is a convergent sequence, given any $\epsilon > 0$, there are a limiting value L and an $N \in \mathbb{Z}^+$ such that $|a_n - L| < \epsilon/2$ whenever $n > N$. So when $m, n > N$, by the triangle inequality,

---

5 Augustin-Louis Cauchy (1789–1857) was a French mathematician who was particularly responsible for bringing exact definitions and rigorous proofs to calculus and analysis.

$$|a_n - a_m| \le |a_n - L| + |L - a_m| < \frac{\epsilon}{2} + \frac{\epsilon}{2} = \epsilon,$$

implying that $\{a_n\}$ is a Cauchy sequence. $\qquad\square$

So every convergent sequence is Cauchy. The question then is, "Is every Cauchy sequence also convergent?" Interestingly, the answer to this question is somewhat complicated. The next chapter will discuss the least upper bound property of the reals, and this property will then be used to show that the reals are *complete*, meaning that Cauchy sequences in the real numbers do always converge to a real number. In settings (spaces) that are not complete, Cauchy sequences need not converge, at least not to an element of the space, but this is a topic beyond the scope of this work. Those wanting to learn about settings (metric spaces) where Cauchy sequences need not converge can look at, for example, Abbott [1, pp. 258–262], Rosenlicht [4, pp. 51–53] or Rudin [5, pp. 252–253].

Although not every Cauchy sequence is convergent, every Cauchy sequence is *bounded* regardless of the space. For sequences of real numbers, this means that there exists a bound, $M > 0$, such that $|a_n| \le M$ for all $n \in \mathbb{Z}^+$.

Theorem 1.8.2. *Suppose that $\{a_n\}$ is a Cauchy sequence. Then $\{a_n\}$ is bounded.*

Proof. Take $\epsilon = 1$. Since $\{a_n\}$ is Cauchy, $\exists N \in \mathbb{Z}^+$ such that $|a_n - a_m| < 1 \; \forall m, n > N$. So taking $m = N + 1$, all elements of the sequence beyond a_N are within 1 of a_{N+1}. Now define

$$M := \max\{|a_1|, |a_2|, \dots, |a_N|, |a_{N+1}| + 1\}.$$

Notice that then $|a_n| \le M$ for all $n \in \mathbb{Z}^+$, meaning that this sequence is bounded. $\qquad\square$

Exercises 1, Group A

1.1. Please compute the limit as $n \to \infty$ for the sequence $\{a_n\}$ for each of the following elements a_n or determine that the limit does not exist (no need to give an ϵ-N proof):

(a) $a_n = \dfrac{3n}{n+1}$

(b) $a_n = \dfrac{n^2 + 2n - 1}{5n^2 + 2}$

(c) $a_n = \dfrac{n^4 - n^2 + 1}{n^3 + n + 1}$

(d) $a_n = \dfrac{3n^2}{2n^3 + 1}$

(e) $a_n = \dfrac{1 + (-1)^n}{2}$

(f) $a_n = \dfrac{1 + (-1)^n}{n}$

(g) $a_n = \dfrac{\sin n}{n}$

(h) $a_n = \dfrac{n}{3}\sin\left(\dfrac{5}{n}\right)$

(i) $a_n = \dfrac{1 + 5^n}{7^n}$

(j) $a_n = \left(1 + \dfrac{1}{2n}\right)^n$

(k) $a_n = \left(\dfrac{3n - 2}{3n + 2}\right)^n$

(l) $a_n = n^{(1/n)}$

(m) $a_n = \dfrac{\ln(n+1)}{n+1}$

(s) $a_n = \dfrac{\sqrt{2n+1} - \sqrt{2n-1}}{e^{-n}}$

(n) $a_n = \dfrac{\ln(n^2+1)}{\ln(2n^2+1)}$

(t) $a_n = \sqrt{n}(\sqrt{n+1} - \sqrt{n})$

(u) $a_n = \dfrac{n!}{n^n}$

(o) $a_n = \dfrac{e^n}{n!}$

(v) $a_n = n^2(1 - \cos(2/n))$

(p) $a_n = \sqrt{n+1} - \sqrt{n-1}$

(w) $a_n = \dfrac{n^2}{n!}$

(q) $a_n = \dfrac{3^n}{n!}$

(x) $a_n = (2/n)^{2/n}$

(r) $a_n = \dfrac{n}{3^n}$

Answers. (b) 1/5, (e) does not exist, (h) 5/3, (j) $\sqrt{e}$, (l) 1, (n) 1, (r) 0, (t) 1/2.

1.2. Consider the sequence $\{1, \pi, 0.5, q, 8, 1, \pi/6, \sqrt{17}, 4, z, 1, 1, 1/99, y, \beta, 1, \ldots\}$. Which of the following would seem to be subsequences of the above sequence?
(a) $\{1, 0.5, q, \pi/6, \sqrt{17}, 4, 1, 1, 1, \ldots\}$
(b) $\{\pi, q, 1, \pi/6, \sqrt{17}, \sqrt{17}, z, 1, 1, 1/99, y, \beta, 1, \ldots\}$
(c) $\{1, 1, 1, 1, 1, \ldots\}$
(d) $\{-1, -\pi, -0.5, -q, -8, -1, -\pi/6, -\sqrt{17}, -4, -z, -1, -1, -1/99, -y, -\beta, -1, \ldots\}$

Answers. (b) No (it appears that $\sqrt{17}$ occurs only once in the original sequence), (c) yes.

1.3. Spot the pattern: For each of the following sequences, please find the value of a_n so that $\{a_n\}$ matches the given sequences. Assume that the first element corresponds to $n = 1$. In some cases, there may be more than one correct answer.

(a) $\{2, 4, 6, 8, 10, \ldots\}$

(f) $\left\{\dfrac{1}{2}, \dfrac{1}{5}, \dfrac{1}{10}, \dfrac{1}{17}, \dfrac{1}{26}, \ldots\right\}$

(b) $\left\{\dfrac{1}{2}, \dfrac{1}{3}, \dfrac{1}{4}, \dfrac{1}{5}, \dfrac{1}{6}, \ldots\right\}$

(g) $\left\{\dfrac{1}{4}, \dfrac{4}{15}, \dfrac{5}{11}, \dfrac{28}{29}, \dfrac{41}{18}, \ldots\right\}$

(c) $\left\{\dfrac{1}{2}, \dfrac{1}{4}, \dfrac{1}{8}, \dfrac{1}{16}, \dfrac{1}{32}, \ldots\right\}$

(h) $\left\{\dfrac{2}{3}, \dfrac{11}{9}, \dfrac{26}{19}, \dfrac{47}{33}, \dfrac{74}{51}, \ldots\right\}$

(d) $\{1, 0, 1, 0, 1, \ldots\}$

(e) $\left\{\dfrac{1}{4}, \dfrac{-1}{9}, \dfrac{1}{14}, \dfrac{-1}{19}, \dfrac{1}{24}, \ldots\right\}$

(i) $\left\{1, \dfrac{7}{17}, \dfrac{11}{53}, \dfrac{19}{161}, \dfrac{7}{97}, \ldots\right\}$

Answers. (b) $a_n = 1/(n+1)$, (d) $a_n = 1 + (-1)^n$, (e) $a_n = (-1)^{n+1}/(5n-1)$, (g) $a_n = (3^{n-1}+1)/(7n+1)$.

Exercises 1, Group B

1.1. Please determine L (limit value) for each of the following sequences $\{a_n\}$ and then show that each sequence satisfies the definition of *limit* by finding N as a function of ε.

(a) $a_n = \dfrac{1}{5n + 2}$

(b) $a_n = \dfrac{1}{n^2 + 1}$

(c) $a_n = \dfrac{1}{n^2 - 4}$

(d) $a_n = \sqrt{n+1} - \sqrt{n}$

(e) $a_n = \dfrac{1 + 2 + 3 + \cdots + n}{n^2}$

Hint: (e) Recall the formula for the sum of the first n integers.

1.2. Please show that the following sequences satisfy the definition of *divergence to infinity* by finding N as a function of M.

(a) $a_n = \dfrac{n^2 + 1}{n - 1}$

(c) $a_n = \dfrac{n}{\sqrt{n+1}}$

(b) $a_n = e^{n\pi}$

(d) $a_n = \dfrac{1}{\sin(1/n)}$

1.3. Please prove the triangle inequality: For any three real numbers a, b and c,

$$|a - b| \le |a - c| + |c - b|.$$

Hint: Show that $|a - c| + |c - b| - |a - b|$ is always nonnegative by considering the six possible orders for a, b and c and then the positivity or negativity of the differences in absolute values. For example, if $a \le b \le c$, then $|a - c| = c - a$.

1.4. Suppose that a sequence $\{a_n\}$ diverges to infinity and that there is a second sequence $\{b_n\}$ where $a_n \le b_n$ for all $n \in \mathbb{Z}^+$. Please show that this second sequence $\{b_n\}$ also diverges to infinity.

1.5. (a) Prove part (b) of the ratio test, that is, prove that if $r > 1$, then $|a_n| \to \infty$.
   Hint: Follow the proof of Theorem 1.4.4 but now with $1 < \rho < r$.
(b) Give examples of sequences where $r = 1$ that either converge or diverge.
   Hint: Suppose a_n is a carefully chosen rational expression in n. Consider Example 1.4.3.

1.6. Use the proof given above that the limit of the sum of two converging sequences must be the sum of the limits of those sequences to show that the limit of the difference of two converging sequences must be the difference of the limits of those sequences:

$$\lim_{n \to \infty} (a_n - b_n) = \left(\lim_{n \to \infty} a_n \right) - \left(\lim_{n \to \infty} b_n \right).$$

1.7. Use item 2 of Theorem 1.5.1 to prove item 3 by replacing B in item 2 with $1/B$ and b_n with $1/b_n$. Notice that by assumption, $B \ne 0$, and since $b_n \to B$, we have $B \ne 0 \implies b_n \ne 0$ at least for sufficiently large n.

1.8. (a) Prove item 4 of Theorem 1.5.1.

    Hint: Notice that $a_n \neq b_n \implies b_n - a_n \geq 0$, so apply item 1.

(b) Give an example to show that even if $a_n < b_n$ in item 4 of Theorem 1.5.1, the conclusion remains $A \leq B$.

    Hint: Find two appropriate sequences that both converge to zero.

1.9. Prove that if the sequence $\{a_n\}$ converges to L, then the sequence $\{a_n^2\}$ converges to L^2. Why is the converse of this statement false? That is, why can a_n^2 converge to L^2 even though a_n does not converge to L? Give a counterexample to support your answer.

    Hint: Suppose a_n alternates in sign between positive and negative.

1.10. Suppose that the sequence $\{a_n\}$ diverges to infinity. Please show that the sequence $\{a_n^2\}$ also diverges to infinity.

1.11. Suppose two sequences $\{a_n\}$ and $\{b_n\}$ both converge to L. Please show that the sequence $\{a_1, b_1, a_2, b_2, a_3, b_3, \ldots\}$ must also converge to L.

1.12. Suppose that the sequence $\{a_n\}$ converges. Show that if a subsequence converges to a limit L, then the entire sequence must also converge to L.

1.13. (a) Please give an example of a sequence that is bounded below and unbounded above and does *not* diverge to infinity.

(b) Please give an example of a sequence that is unbounded below and unbounded above and does not diverge to infinity.

(c) Please show that any sequence that is unbounded below cannot diverge to infinity.

1.14. Give an example of a real sequence $\{a_n\}$ with $a_n > 0$ such that $\{a_n\}$ and $\{na_n\}$ both converge to the same limit.

    Hint: What must L be?

1.15. Please show that the sequence $\{1/n^2\}$ is Cauchy convergent.

1.16. Please show that any subsequence of a Cauchy sequence is itself Cauchy.

2 The real numbers, $\mathbb{R}$

Before we can discuss sequence convergence further, or start to discuss series, we need to consider a few basic properties of the real numbers, $\mathbb{R}$. Perhaps surprisingly, some of these basic mathematical facts are not widely known, sometimes even among those who use calculus routinely. This chapter begins with a discussion of the axioms that identify the real numbers, then considers some of the implications of these axioms, then considers how many numbers of different types there are, and finally closes with some basic results in logic.

2.1 Axioms of the real numbers

What are the real numbers? Most students start calculus with an intuitive understanding of what the real numbers are, but few understand them from a mathematical point of view.

The modern mathematical understanding of the real numbers comes from the late nineteenth century—the work of Georg Cantor[1] and others. It is based on identifying the fundamental properties or axioms that separate the reals from other types of numbers (for example, the integers, $\mathbb{Z}$, the rationals, $\mathbb{Q}$, and the complex numbers, $\mathbb{C}$). These axioms are normally divided into three sets: the field axioms, the order axiom and the least upper bound axiom.

The field axioms define how the basic operations of addition and multiplication work for the reals (and any other field). They give what are generally seen as obvious properties, but they are also fundamental properties in the sense that all other similar properties can be obtained from these axioms. The field axioms are given in Table 2.1. These six axioms are satisfied by all real numbers (except that there is no multiplicative inverse for 0). Of course, they are also satisfied by any other field, including both the rationals and the complex numbers. A familiar set of numbers that is *not* a field is the integers. They indeed satisfy most of the field axioms, but they do not generally have multiplicative inverses. For example, $2 \in \mathbb{Z}$, but $1/2 \notin \mathbb{Z}$.

As mentioned above, the field axioms are fundamental because they allow us to establish all of the other familiar arithmetic properties of the rationals, the reals and the complex numbers. The next example shows how one can use the field axioms to establish the familiar property of additive cancelation that must hold in any field.

Example 2.1.1. Use the field axioms to show that any field must have additive cancelation: If a, b and x are real numbers, then

$$a + x = b + x \quad \Longrightarrow \quad a = b.$$

---

[1] Georg Cantor (1845–1918) was a German mathematician who is famous for theorems and counterexamples that disprove results that were widely expected to be true, making him rather controversial in his lifetime.

https://doi.org/10.1515/9783110768398-002

Table 2.1: Suppose that x, y and z are *any* real numbers **except** for the multiplicative inverse where $x \neq 0$ (the zero has no multiplicative inverse). Then the six axioms (11 statements) above must hold for the real numbers and indeed for any field. For example, the closure axiom for addition states that the sum of any two real numbers is also a real number.

Field axioms	Given $x, y, z \in \mathbb{R}$	
	Addition	Multiplication
Closure	$x + y \in \mathbb{R}$	$xy \in \mathbb{R}$
Commutativity	$x + y = y + x$	$xy = yx$
Associativity	$(x + y) + z = x + (y + z)$	$(xy)z = x(yz)$
Identities	$\exists 0 \in \mathbb{R}, x + 0 = x$	$\exists 1 \in \mathbb{R}, x(1) = x$
Inverses	$\exists (-x) \in \mathbb{R}, x + (-x) = 0$	For $x \neq 0, \exists(1/x) \in \mathbb{R}, x(1/x) = 1$
Distributivity	$x(y + z) = xy + xz$	

Answer. This type of result can often be given as a one-line proof, starting at a and ending with b, but perhaps it is clearer to spread it out:

$$
\begin{aligned}
a &= a + 0 && \text{identity axiom} \\
&= a + (x + (-x)) && \text{inverse axiom} \\
&= (a + x) + (-x) && \text{associativity} \\
&= (b + x) + (-x) && \text{given assumption} \\
&= b + (x + (-x)) && \text{associativity} \\
&= b + 0 && \text{inverse axiom} \\
&= b. && \text{identity axiom}
\end{aligned}
$$

There are a number of similar basic results that follow from the field axioms; a couple are discussed in Exercise B.2.1.

There now need to be additional axioms to distinguish the reals from the rationals and the complex numbers. The next axiom to consider, the order axiom, will separate the real numbers from the complex numbers; this axiom is given in Table 2.2.

One might think that the order axiom just states obvious properties that every type of number satisfies, but this is not the case. In particular, the complex numbers *cannot* be

Table 2.2: Suppose that x, y and z are *any* real numbers and that the symbol < (less than) has its usual meaning, that is, $x < y$ if and only if $y - x$ is positive. Then the three clauses above form an axiom that must hold for the real numbers and indeed for any ordered field. For trichotomy, the three cases are mutually exclusive. Trichotomy can also be stated as $y - x \in \mathbb{R}_+, y - x = 0$ or $x - y = -(y - x) \in \mathbb{R}_+$.

Order axiom	$\exists \mathbb{R}_+ \subset \mathbb{R}$ (positive reals) such that
Addition	If $x, y \in \mathbb{R}_+$, then $x + y \in \mathbb{R}_+$
Multiplication	If $x, y \in \mathbb{R}_+$, then $xy \in \mathbb{R}_+$
Trichotomy	For $x, y \in \mathbb{R}$, we have $x < y, x = y$ or $y < x$

ordered. To see this, suppose that $0 < i := \sqrt{-1}$. Then, if one multiplies both sides of $0 < i$ by the supposed positive number i, one obtains $0 < -1$, which is a problematic result (to say the least). If one now multiplies both sides of $0 < -1$ by i, one obtains $0 < -i$, which violates the principle of trichotomy. Starting with $i < 0$ leads to similar contradictory results (see Exercise B.2.2), and since $i = 0$ is not possible either, trichotomy (among other parts of order) cannot be satisfied.

So the integers are not a field, and the complex numbers cannot be ordered; we are left with the task of separating the reals and the rationals. The final axiom for the real numbers does this by describing the key property of the reals that is not shared by the rationals: completeness. This **least upper bound** (LUB) axiom is given in Table 2.3.

Table 2.3: That the least upper bound of a set is itself a real number is a key part of the least upper bound axiom. Why does this property not hold for ℚ?

Least upper bound axiom	
Least upper bound	Every nonempty subset $S \subset \mathbb{R}$ that is bounded above has a least upper bound, LUB(S), that is itself a real number.

If the least upper bound is the smallest possible upper bound, what is an upper bound?

Definition. Suppose that $S \subset \mathbb{R}$ and that there is an $M \in \mathbb{R}$ such that $x \leq M \; \forall x \in S$. Then M is an **upper bound** for S, and S is **bounded above** by M. Similarly suppose that there is an $m \in \mathbb{R}$ such that $m \leq x \; \forall x \in S$. Then m is a **lower bound** for S, and S is **bounded below** by m.

Perhaps surprisingly, this final axiom is often seen as the key element that makes the real numbers special (though logically it is no more fundamental than any of the other axioms). The next section will explore the concept of least upper bound at some length, but before we turn to this, it is worth noting what all of the axioms together imply. The real numbers are *the unique complete ordered field*. No other entity satisfies the field axioms, the order axiom and the least upper bound axiom. We could use different names for each real number (we could say their names in Dutch, Arabic, Mandarin, Swahili or any other language), but the numbers themselves are basically the same. They are the unique entity that satisfies all the above axioms.

Finally, it is possible to flip the entire set about zero to achieve a greatest lower bound of a set: Every nonempty subset $S \subset \mathbb{R}$ that is bounded below has a greatest lower bound, GLB(S), that is also a real number. Notice that a set having a greatest lower bound is equivalent to having a least upper bound since GLB(S) = −LUB(−S), where

$$-S := \{x \in \mathbb{R} \mid -x \in S\}$$

(see Exercise B.2.4).

2.2 Completeness and the least upper bound

Although it is not obvious, the completeness of the real numbers and the least upper bound property of the real numbers are actually the same thing. This discussion will not cover the concept of completeness for the real numbers in detail, but we will attempt to help the reader have a better sense of what completeness and the least upper bound property imply for the reals.

Example 2.2.1. Consider the sequence

$$S = \left\{ 3, \frac{31}{10}, \frac{314}{100}, \frac{3141}{1000}, \frac{31415}{10000} \cdots \right\}.$$

This is pretty clearly a sequence of *rational* numbers (each element is the ratio of integers). Does this sequence converge to anything?

Answer. Most readers, if they think for a moment, will agree that (assuming the rest of the elements of the sequence continue to follow the pattern that the first few elements suggest) this sequence converges to π. But most readers likely also know that $\pi \notin \mathbb{Q}$ (π is not a rational number). So an increasing sequence of rational numbers that is bounded above does not have to converge to a rational number.

Of course the elements of the above sequence are also real numbers, as is π. This gets at the important property that the reals are complete: convergent sequences of real numbers converge to real numbers, never to something outside the reals. The rationals simply do not have this sort of property. But notice something else: If the above sequence is thought of as a set of real numbers, then because this set is both nonempty and bounded above (by, say, 4), the least upper bound axiom asserts that this set must have a *least* upper bound. Notice that because this sequence is increasing and converging to π, $\pi = \text{LUB}(S)$.

There is *much* more that can be said about completeness, but for now, there are only two rigorous results based on the least upper bound axiom to be proven here. Both of these results depend on the definition of a special set called S in each case.

Theorem 2.2.1. *Any Cauchy sequence of real numbers converges to a real number. Hence the real numbers are complete.*

Proof. Suppose that $\{a_n\}$ is a Cauchy sequences of real numbers. Define

$$S := \{x \in \mathbb{R} \mid x \leq a_n \text{ for an infinite number of } n \in \mathbb{Z}^+\}.$$

By Theorem 1.8.2, every Cauchy sequence is bounded, so S must be bounded above (elements of $\{a_n\}$ cannot grow towards $+\infty$), and S must be nonempty (elements of $\{a_n\}$ cannot head towards $-\infty$). By the least upper bound property of the real numbers, S must have a least upper bound. Let $a := \text{LUB}(S)$; we must show that $\{a_n\}$ converges to a.

Given $\epsilon > 0$, since $\{a_n\}$ is Cauchy, $\exists N \in \mathbb{Z}^+$ such that $|a_n - a_m| < \epsilon/2 \, \forall \, m, n > N$. Now only a finite number of elements of the sequence can be greater than $a + \epsilon/2$, for if not, a is not an upper bound for S. Also there are infinitely many elements of the sequence above $a - \epsilon/2$, for if not, $a - \epsilon/2$ would be an upper bound for S less than $a = \mathrm{LUB}(S)$. Hence a specific integer $m^* > N$ can be chosen such that $a - \epsilon/2 < a_{m^*} \leq a + \epsilon/2$, implying that $|a_{m^*} - a| \leq \epsilon/2$. By the triangle inequality, then,

$$|a_n - a| \leq |a_n - a_{m^*}| + |a_{m^*} - a| < \epsilon/2 + \epsilon/2 = \epsilon$$

$\forall \, n > N$. Therefore by definition $\{a_n\}$ converges to a. $\qquad\qquad\qquad\qquad\qquad\square$

Before proving the second result, an exact definition for the terms *increasing* and *decreasing* is needed:

Definition. A sequence of real numbers $\{a_n\}$ is **increasing** iff $a_n \leq a_{n+1} \, \forall \, n \in \mathbb{Z}^+$. Similarly, a sequence of real numbers $\{b_n\}$ is **decreasing** iff $b_n \geq b_{n+1} \, \forall \, n \in \mathbb{Z}^+$.

Remark. Using these definitions, a constant sequence is both increasing and decreasing. This is indeed unfortunate, but nonetheless, this definition works best for the following result.

Theorem 2.2.2 (Monotone convergence theorem). *An increasing sequence of real numbers $\{a_n\}$ that is bounded above must converge to its least upper bound. Similarly, a decreasing sequence of real numbers $\{b_n\}$ that is bounded below must converge to its greatest lower bound.*

Remark. Both of the previous two theorems are important because they allow us to show that certain sequences converge without resorting to the definition. In particular, while the value of the limit L must be known in order to directly apply the definition to a specific sequences, these theorems do not require that L be known. Increasing sequences can be shown to converge without having to first find L if a bound can be found. While the proof of this theorem will use the definition, its applications will not.

Proof. Since $\{a_n\}$ is bounded above, there exists $M \in \mathbb{R}$ such that $a_n \leq M \, \forall \, n \in \mathbb{Z}^+$. Let S be the *set* corresponding to this sequence:

$$S := \{x \in \mathbb{R} \mid x = a_n \text{ for some } n \in \mathbb{Z}^+\}.$$

Since S is a nonempty set that is bounded above, by the least upper bound axiom, there exists $L := \mathrm{LUB}(S)$. Now given $\epsilon > 0$, there exists $N \in \mathbb{Z}^+$ such that $L - \epsilon < a_N$ (for if not, then $L - \epsilon$ is an upper bound less than the least upper bound L). But $a_n > a_N \, \forall \, n > N$ since the sequence is increasing, and hence

$$|a_n - L| = L - a_n \leq L - a_N < \epsilon$$

$\forall \, n > N$. Thus, $\{a_n\}$ converges to L.

The proof of convergence when the sequence $\{b_n\}$ is bounded below and monotonically decreasing is based on considering the sequence $\{-b_n\}$, which must be monotonically increasing. See Exercise B.2.5. ☐

Now we can put the previous theorem to good use on a type of sequence we have not studied up until now.

Example 2.2.2. Consider the sequence

$$\left\{ \sqrt{2},\ \sqrt{2 + \sqrt{2}},\ \sqrt{2 + \sqrt{2 + \sqrt{2}}}, \dots \right\}.$$

Does this sequence converge, and if so, to what value?

Answer. What is a_n (the n-th element) for this sequence? Notice that it is rather difficult to write down a closed-form formula for the n-th element, so trying to find a_n is probably not the best way to proceed. Instead we will show that this sequence converges and find its limit without ever writing down a_n explicitly.

Most people who have not studied this sort of sequence before cannot immediately see why it must converge or what the value of its limit will be. Often the key observation for a sequence when there is no easy formula for the n-th element is to notice that a recurrence relation is possible. A **recurrence relation** is an expression giving the n-th element in terms of the $(n - 1)$-th element. For the sequence at hand, the recurrence relation is just

$$a_n = \sqrt{2 + a_{n-1}}.$$

Generally speaking, a recurrence relation is not as good as a closed-form formula for the n-th element, but it is often enough to show that a given sequence is both *increasing* and *bounded above*. We can prove both of these through induction.

In its most basic form, proof by induction proceeds through a two-step process: First one shows that the desired statement is true for $n = 1$. This first step is often easy, maybe even obvious, but it is certainly essential. The second step is to *assume* that the statement is true for $n - 1$ and then *prove* that the statement is true for n using the truth of the statement for $n - 1$. The point is that since the $n = 1$ statement has been proven true, because of the second step, the $n = 1$ statement implies the truth of the $n = 2$ statement. But then the $n = 2$ statement implies the truth of the $n = 3$ statement, and so forth. This gives us the result that the statement is true for all $n \in \mathbb{Z}^+$.

To see that this current sequence is increasing, notice that since the square root function is increasing, $a_1 = \sqrt{2} = \sqrt{2 + 0} < \sqrt{2 + \sqrt{2}} = a_2$. So the statement $a_n < a_{n+1}$ is true at least when $n = 1$. Next one must assume that $a_{n-1} < a_n$ and use this to show that $a_n < a_{n+1}$. To do this, the recurrence relation is essential. Consider the difference

$$a_{n+1}^2 - a_n^2 = (2 + a_n) - (2 + a_{n-1})$$
$$= a_n - a_{n-1} > 0,$$

where the inequality above follows from the inductive assumption that $a_{n-1} < a_n$. Since each element of the sequence is positive,

$$a_{n+1}^2 - a_n^2 > 0 \quad \Longrightarrow \quad a_{n+1} > a_n.$$

Remark. It is important to keep in mind that the formal inductive proof is not the only way to prove many results. For example, one can prove that the current sequence is increasing directly without using induction (see Exercise B.2.6).

Now to see that this sequence is bounded above, notice that $a_n < 4 \, \forall \, n \in \mathbb{Z}^+$ because (1) $a_1 = \sqrt{2} < 4$ and (2) assuming that $a_{n-1} < 4$, one sees that

$$a_n = \sqrt{2 + a_{n-1}} < \sqrt{2 + 4} < \sqrt{16} = 4$$

due to the recurrence relation. Again notice that the argument uses the fact that the square root function is increasing and the useful if simple fact that $2 + 4 = 6 < 16$. Thus, since the sequence is both increasing and bounded above, by the monotone convergence theorem, our sequence must converge.

Now we consider a separate but related question: What does our sequence converge to? The answer to this question again follows from the recurrence relation: If there exists $L \in \mathbb{R}$ such that $a_n \to L$, then also $a_{n-1} \to L$. So in the limit, the recurrence relation becomes

$$L = \sqrt{2 + L},$$

or, because each element of the sequence and L are positive,

$$L^2 - L - 2 = 0.$$

Solving this quadratic equation, one finds that $L = 2$.

Remark. It might come as a bit of a surprise that this rather strange sequence converges to a common number, even an integer. From a mathematical point of view, this result should not be a surprise: some rather simple looking sequences converge to complicated numbers, while some more complicated sequences like the current one converge to simple numbers.

2.3 How many reals are there?

How many reals are there? For that matter, how many rationals are there? Certainly it is true that

$$\mathbb{Z} \subset \mathbb{Q} \subset \mathbb{R},$$

and these are proper subsets (there are real numbers that are not rationals and rational numbers that are not integers). So one might suspect that the number of reals is greater than the number of rationals and that the number of rationals is greater than the number of integers. On the other hand, most people would agree intuitively that there are an infinite number of integers, rationals and reals. So maybe the answer should be that all three sets have the same number of elements. All of this vague discussion should make one long for a clear mathematical view, and that is what we work through next.

The number of members of any set is its **cardinality**. For finite sets, the cardinality is what you think it is. For example, the cardinality of the set $\{9, \pi, t, \text{house}\}$ is 4, even if this is a rather strange, eclectic set. For infinite sets (sets that have more than a finite number of elements), the definitions are more involved.

Definition. A set is **countable** or **countably infinite** iff it can be put in one-to-one correspondence with the positive integers.

Theorem 2.3.1. *The integers are countable.*

Proof. Obviously the positive integers are themselves countable, because each can be assigned to itself in a one-to-one correspondence:

$$1 \longleftrightarrow 1$$
$$2 \longleftrightarrow 2$$
$$3 \longleftrightarrow 3$$
$$\vdots \quad \vdots \quad \vdots$$
$$n \longleftrightarrow n$$
$$\vdots \quad \vdots \quad \vdots$$

What might be less obvious is the fact that we can do the same thing for *all* integers:

$$1 \longleftrightarrow 0$$
$$2 \longleftrightarrow 1$$
$$3 \longleftrightarrow -1$$
$$4 \longleftrightarrow 2$$
$$5 \longleftrightarrow -2$$
$$\vdots \quad \vdots \quad \vdots$$
$$2n \longleftrightarrow n$$
$$2n + 1 \longleftrightarrow -n$$
$$\vdots \quad \vdots \quad \vdots$$

This means that in our count, 0 is the first integer, 1 is the second integer, −1 is the third integer, and so forth. Notice that we can immediately say that, for example, the 37th integer in our count is −18 (37 = 2n + 1 $\implies$ n = 18 $\implies$ −n = −18) and that 142 is the 284th integer in our count. The key to including all the integers in this count is the alternation between positive and negative integers. □

Thus, the set of all the integers, ℤ, is countable, and thus there are exactly the same number of integers as positive integers, even though there are integers that are not positive integers. What about the rational numbers? Are the rationals countable, or are there too many rationals to count? Perhaps surprisingly, the rationals *are* countable. The next result gives a counting of the positive rationals; the counting of all rationals is left as an exercise (Exercise A.2.3).

Theorem 2.3.2. *The positive rationals are countable.*

Proof. Making this count requires a scheme that allows us to move through all the positive rationals. The key here is putting the positive rationals in an array such as the one in Figure 2.1 and then moving systematically through them. Notice that if we started by counting across any row or down any column (or diagonally downward and to the right), we would never return to count the numbers that we were passing in the rows above or below, or the columns to the right or the left. So what is needed is a pattern that eventually reaches every number (every entry in the array). Perhaps the easiest such pattern is to count downward and to the *left* from the top row along diagonals were $p + q$ is constant (p-values are in the top row; q-values are in the left column). Each time such a diagonal is fully counted, we move to the next value of $p + q$ and count the next diagonal. But notice that to avoid counting some rationals multiple times, we must skip over rationals that have already been counted: $2/1 = 4/2 = 6/3$, for example. So our counting is

$$1 \longleftrightarrow 1/1 = 1$$
$$2 \longleftrightarrow 2/1 = 2$$
$$3 \longleftrightarrow 1/2$$
$$4 \longleftrightarrow 3/1 = 3$$
$$5 \longleftrightarrow 1/3$$
$$6 \longleftrightarrow 4/1 = 4$$
$$7 \longleftrightarrow 3/2$$
$$8 \longleftrightarrow 2/3$$
$$9 \longleftrightarrow 1/4$$
$$10 \longleftrightarrow 5/1 = 5$$
$$11 \longleftrightarrow 1/5$$
$$\vdots \quad \vdots \quad \vdots$$

p/q	1	2	3	4	5	…
1	1/1	2/1	3/1	4/1	5/1	…
2	1/2	2/2	3/2	4/2	5/2	…
3	1/3	2/3	3/3	4/3	5/3	…
4	1/4	2/4	3/4	4/4	5/4	…
5	1/5	2/5	3/5	4/5	5/5	…
⋮	⋮	⋮	⋮	⋮	⋮	

Figure 2.1: An array of the positive rational numbers. The rational number p/q is in the p-th column and the q-th row. The color-coded boxes indicate which rationals (when written in lowest terms) have a certain $p + q$ value: for yellow, $p + q = 2$; for blue, $p + q = 3$; for red, $p + q = 4$; for green, $p + q = 5$; for orange, $p + q = 6$. For each $p + q$ value, the count proceeds from the top downward to the left; then it moves to the next $p + q$ value.

Notice that the distance between integers generally increases as the value of $p + q$ increases and that this counting is guaranteed to include each positive rational exactly once, and thus shows that the positive rationals are countable. □

Remark. The counting of $\mathbb{Q}^+$ presented in the proof of Theorem 2.3.2 is a fairly basic one. Much more elegant countings are possible; see for example Graver [3].

Our success so far in counting the rationals might lead one to think that somehow any infinite set is countable. This is not the case, as the next example shows.

Theorem 2.3.3 (Cantor). *The reals are **not** countable.*

Proof. Showing that a set is not countable is a rather different thing from showing that a set is countable. This time we must proceed by contradiction: we assume that the reals *are* countable and show that this leads to an unavoidable contradiction.

Consider the interval $[0, 1] \subset \mathbb{R}$ and suppose that $\{x_i\}$ is an enumeration (a counting) of the reals in this interval (if all the reals are countable, then surely those in $[0, 1]$ are too). Let us write this enumeration in decimal form so that $x_i = 0.x_{i1}x_{i2}x_{i3}x_{i4}x_{i5}\ldots$, where $x_{ij} \in \{0, 1, 2, 3, 4, 5, 6, 7, 8, 9\}$ are the digits of our decimals. To avoid representing certain reals twice, we must agree that terminating decimals (those whose expansions end in all zeros) will only be represented in this format, not by the alternative where the expansion ends in all nines. For example, 1/2 is represented as $0.500000\ldots$, not as $0.499999\ldots$.[2]

2 Readers not familiar with this issue should consider Example 3.2.4 in the section below on geometric series.

Now consider our enumeration as written in the array in Figure 2.2. There is at least one real number that is not in our enumeration: $a := 0.a_1a_2a_3a_4a_5 \ldots$, where $a_i := x_{ii} + 5 \pmod{10}$. That is, define a_i by adding 5 to x_{ii}, and if the result is 10 or more, subtract 10 to get a single digit. So for example, if $x_{ii} = 3$, then $a_i = 8$, while if $x_{ii} = 7$, then $a_i = 2$. Thus, a differs from the first entry in our enumeration x_1 because they differ by five digits in the first decimal place; a differs from the second entry x_2 because they differ by five digits in the second decimal place; and so forth. Our enumeration is therefore missing a number no matter what enumeration we chose. □

$$
\begin{aligned}
x_1 &= 0.x_{11}x_{12}x_{13}x_{14}x_{15}\cdots \\
x_2 &= 0.x_{21}x_{22}x_{23}x_{24}x_{25}\cdots \\
x_3 &= 0.x_{31}x_{32}x_{33}x_{34}x_{35}\cdots \\
x_4 &= 0.x_{41}x_{42}x_{43}x_{44}x_{45}\cdots \\
x_5 &= 0.x_{51}x_{52}x_{53}x_{54}x_{55}\cdots \\
&\vdots
\end{aligned}
$$

Figure 2.2: A supposed enumeration of the real numbers in the interval $[0, 1]$. Each $x_{ij} \in \{0, 1, 2, 3, 4, 5, 6, 7, 8, 9\}$. There is at least one real $a = 0.a_1a_2a_3a_4a_5 \ldots$ in the interval missing from this list; choose a to differ from the entries on the list at the entries on the main diagonal boxed in red.

The material in this section (the countability of the rationals and the uncountability of the reals) is perhaps the most fundamental fact about nature that most people—even well-educated people—really do not know. These arguments come from the late nineteenth century; the final piece (that the reals are uncountable) is due to Cantor. It is also true, however, that many mathematicians were surprised, even unhappy, when Cantor presented this result.

2.4 A bit of logic

This section does not deal specifically with the real numbers, but rather with a bit of logic that is fundamental to the mathematics of calculus and the real numbers. The key concept discussed here is the *contrapositive*.

Consider statements of the form $A \implies B$ (A implies B), also called if-then statements or implications. There are many examples of statements in this format; for example,

if x is a rational number, then x is a real number

is such a statement. It can also be written symbolically:

$$x \in \mathbb{Q} \implies x \in \mathbb{R}.$$

The part of the statement following the "if" is called the **hypothesis**; the part following the "then" is called the **conclusion**. So in this example, the hypothesis is $x \in \mathbb{Q}$ and the conclusion is $x \in \mathbb{R}$.

The **contrapositive** of the implication $A \implies B$ is the statement obtained by transposing A and B and negating both. So the contrapositive of "A implies B" is "not B implies not A." In symbols, the contrapositive is $\neg B \implies \neg A$, that is, "not" is symbolized by "$\neg$." Again, returning to the example, the contrapositive of

$$\text{if } x \text{ is a rational number, then } x \text{ is a real number}$$

is

$$\text{if } x \text{ is not a real number, then } x \text{ is not a rational number,}$$

or in symbols, $x \notin \mathbb{R} \implies x \notin \mathbb{Q}$.

The important mathematical (logical) principle is that a statement and its contrapositive are equivalent: they are either both true or both false. Consider again the example: $x \in \mathbb{Q} \implies x \in \mathbb{R}$ is equivalent to $x \notin \mathbb{R} \implies x \notin \mathbb{Q}$; if x is rational, it must be real, so if it is not real, it cannot be rational.

There are three other words that arise in discussions involving if-then statements: converse, inverse and negation. For the statement $A \implies B$, the **converse** is $B \implies A$; notice that the converse is *not* equivalent to the original statement. For example, $x \in \mathbb{Q} \implies x \in \mathbb{R}$ is true, while $x \in \mathbb{R} \implies x \in \mathbb{Q}$ is false. Many real numbers are irrational. What is equivalent to the converse is the **inverse**, which is the contrapositive of the converse. So for $A \implies B$, the inverse is $\neg A \implies \neg B$. Again the inverse is equivalent to the converse, and different from the original statement. The **negation** of the statement $A \implies B$ is $A \wedge \neg B$, meaning that A and the opposite of B exist together. So for example, $x \in \mathbb{R} \implies x \in \mathbb{Q}$ is definitely false, while $x \in \mathbb{R} \wedge x \notin \mathbb{Q}$ (x is real and x is not rational) is true since now x can be irrational.

Finally, let us consider the negation of more general statements, not just implications. In general, one expects to see "$\exists$" and "$\forall$" (the existential and universal quantifiers) get interchanged, "and" and "or" conjunctions get interchanged, and sometimes the order of the statement (at least in English) gets reversed. For example, the negation of

$$\text{for all } x \in G, \text{ there exists } \epsilon > 0 \text{ such that } \{\xi \in \mathbb{R} \mid |\xi - x| < \epsilon\} \subset G$$

is

$$\text{there exists } x \in G \text{ such that, for all } \epsilon > 0, \{\xi \in \mathbb{R} \mid |\xi - x| < \epsilon\} \not\subset G.$$

The set $\{\xi \in \mathbb{R} \mid |\xi - x| < \epsilon\}$ is described in words as "the set of all real numbers ξ such that the absolute difference of ξ and x is less than ϵ." Notice that there is only one negation (the "$\not\subset$"), and it appears deep in the statement. The original statement says that G is *open*, while its negation (the second statement) says that G is not open.

Example 2.4.3. For each of the following implications, please give its contrapositive, its negation, its converse and its inverse. Which statements are true, and which are false?
1. If $x < 1$, then $x \in \mathbb{Z}$.
2. When we get to Paris, we'll be near the Eiffel Tower.

Answer.
1. **Contrapositive**: If $x \notin \mathbb{Z}$, then $x \geq 1$.
   Converse: If $x \in \mathbb{Z}$, then $x < 1$.
   Inverse: If $x \geq 1$, then $x \notin \mathbb{Z}$.
   Negation: There are $x < 1$ that are not integers.
   The original statement, its contrapositive, its converse and its inverse are all false (that the statement is false immediately implies that its contrapositive must be false). The negation, on the other hand, must be true.
2. **Contrapositive**: If we're not near the Eiffel Tower, we won't have gotten to Paris.
   Converse: If we're near the Eiffel Tower, then we'll have arrived in Paris.
   Inverse: If we haven't gotten to Paris, we won't be near the Eiffel Tower.
   Negation: We can be in Paris and not be near the Eiffel Tower.
   The veracity of this statement depends on the exact meaning of "near." If we agree that all of Paris is near the Eiffel Tower (relative to most other points on earth), then the original statement and its contrapositive are true. As for the converse and the inverse, since there are parts of Puteaux (which is separate from Paris) that are closer to the Eiffel Tower than points on the eastern side of Paris, both of these are arguably false. If we agree that the statement is true, then the negation must be false.

Exercises 2, Group A

2.1. Construct a sequence which converges to e (the base of the natural logarithm) similar to the sequence in Example 2.2.1.

2.2. Construct a sequence whose least upper bound is 7, ...
(a) ... and which converges to 7.
(b) ... but which does not converge to 7.

2.3. Show that the rationals are countable.
   Hint: Combine the approaches in Theorem 2.3.1 and Theorem 2.3.2. Where should one count zero?

2.4. For each of the following implications, please give its contrapositive, its negation, its converse and its inverse. Which statements are true, and which are false?
(a) If $x \in \mathbb{Q}$, then $x + 1 \in \mathbb{Q}$.
(b) If $x < 0$, then $\sqrt{x} \in \mathbb{R}$.

(c) If c is the length of the hypotenuse of a right triangle, and a and b are the lengths of the other two sides, then $a^2 + b^2 = c^2$.

(d) If $\sqrt{x} \in \mathbb{Z}^+$, then x is a perfect square.

(e) If $y = a + b$ when a and b are both positive, then $y = a - b$.

Exercises 2, Group B

2.1. (a) Suppose that $x \neq 0$. Use the field axioms to show that any field must have multiplicative cancelation for such an x:

$$ax = bx \quad \Longrightarrow \quad a = b.$$

Hint: Consider Example 2.1.1.

(b) Notice that the statement

$$ax = bx \quad \Longrightarrow \quad a = b$$

fails to be true if $x = 0$. Which part of your argument in (a) fails when $x = 0$?

2.2. Show that $i < 0$ violates the order axiom (where $i := \sqrt{-1}$).

2.3. Please prove that the order axiom for the reals implies **transitivity**: Given that $x, y, z \in \mathbb{R}$, if $x < y$ and $y < z$, then $x < z$.

Hint: Recall that by definition $x < y$ is equivalent to $y - x$ being positive and that the addition clause of the order axiom implies that the sum of two positive real numbers is also positive.

2.4. Prove that if S is a bounded subset of the real numbers ($S \subset \mathbb{R}$), then $\mathrm{GLB}(S) = -\mathrm{LUB}(-S)$, where

$$-S := \{x \in \mathbb{R} \mid -x \in S\}$$

and $\mathrm{GLB}(S)$ is a lower bound of S that is at least as large as any other lower bound.

2.5. Please show that a decreasing sequence $\{b_n\}$ that is bounded below must converge to its greatest lower bound.

Hint: Explain why the sequence $\{-b_n\}$ is increasing and bounded above, and therefore that one can apply the first portion of the monotone convergence theorem.

2.6. Show that the sequence in Example 2.2.2 is increasing through the following two steps:

(a) Prove that $\sqrt{2} < \sqrt{2 + \sqrt{2}}$.

Hint: Show that the square root function is increasing by showing that its derivative is positive. Then notice that $2 < 2 + \sqrt{2}$.

(b) Prove that two consecutive elements of this sequence are

$$\sqrt{2 + \cdots + \sqrt{2}} \quad \text{and} \quad \sqrt{2 + \cdots + \sqrt{2 + \sqrt{2}}}$$

and explain why this implies that the sequence is increasing.

2.7. For the sequence

$$\left\{ \sqrt{3}, \sqrt{3 + \sqrt{3}}, \sqrt{3 + \sqrt{3 + \sqrt{3}}}, \ldots \right\},$$

why must this sequence converge and what is its limit?

2.8. Explain why the real numbers ℝ cannot be written as the countable union of countable sets.

3 Infinite series in $\mathbb{R}$

One of the reasons for studying sequences in the context of calculus is that they are needed to define and study series. The notation for a series is

$$\sum_{k=1}^{\infty} a_k,$$

where, as was the case for our sequences, $a_k \in \mathbb{R}$. As this notation suggests, the idea of a series is that it is a sort of infinite sum. There is, however, a problem with this view: What does it mean to add together an infinite number of real numbers? That it is difficult (or maybe impossible) to work directly with an infinite sum leads to infinite series being defined as something that we do know how to work with—as the limit of a sequence.

Definition. Consider the sequence $\{a_k\} \subset \mathbb{R}$. The **partial sums** for this sequence are defined as

$$A_n := \sum_{k=1}^{n} a_k \equiv a_1 + a_2 + a_3 + \cdots + a_n,$$

that is, A_n is the *finite* sum of the first n elements of the sequence $\{a_k\}$. Notice that $\{A_n\}$ is itself a sequence. An **infinite series** (or simply a **series**) of real numbers is defined as the limit of the sequence of partial sums:

$$\sum_{k=1}^{\infty} a_k := \lim_{n \to \infty} A_n \equiv \lim_{n \to \infty} \sum_{k=1}^{n} a_k,$$

where again $a_k \in \mathbb{R}$. A series **converges** or **diverges** as its sequence of partial sums converges or diverges.

Remarks.
1. Almost every result we prove below about series uses this definition of series as the limit of the sequence of partial sums. This is the key definition for our work in this chapter.
2. In calculus, the most important series are series of functions including power series. For this chapter, however, we will only consider series where the terms a_k are real numbers. Series of functions and power series are discussed in Chapter 4.

The approach for the rest of this chapter is to study series of real numbers, beginning with the most specific and then moving to more general series.

3.1 Harmonic series

Perhaps surprisingly, one of the most important series *does not converge.*

https://doi.org/10.1515/9783110768398-003

Definition. The following specific series is the **harmonic series:**

$$\sum_{k=1}^{\infty} \frac{1}{k} := \lim_{n\to\infty} \sum_{k=1}^{n} \frac{1}{k} \equiv \lim_{n\to\infty}\left(1 + \frac{1}{2} + \frac{1}{3} + \cdots + \frac{1}{n}\right).$$

Theorem 3.1.1. *The harmonic series diverges.*

Proof. To see that the harmonic series diverges, consider the partial sums:

$$A_n := \sum_{k=1}^{n} \frac{1}{k}.$$

Then several values A_n tell the tale:

$$A_1 = 1 \qquad\qquad\qquad\qquad\qquad\qquad\qquad\qquad\qquad > 1\left(\frac{1}{2}\right),$$

$$A_2 = 1 + \frac{1}{2} \qquad\qquad\qquad\qquad\qquad\qquad\qquad\qquad > 2\left(\frac{1}{2}\right),$$

$$A_4 = 1 + \frac{1}{2} + \frac{1}{3} + \frac{1}{4} \qquad\qquad\qquad\qquad\qquad > 3\left(\frac{1}{2}\right),$$

$$A_8 = 1 + \frac{1}{2} + \frac{1}{3} + \frac{1}{4} + \frac{1}{5} + \frac{1}{6} + \frac{1}{7} + \frac{1}{8} \qquad > 4\left(\frac{1}{2}\right),$$

$$A_{16} = 1 + \frac{1}{2} + \frac{1}{3} + \frac{1}{4} + \frac{1}{5} + \frac{1}{6} + \frac{1}{7} + \frac{1}{8} + \frac{1}{9} + \frac{1}{10} + \frac{1}{11} + \frac{1}{12} + \frac{1}{13} + \frac{1}{14} + \frac{1}{15} + \frac{1}{16} > 5\left(\frac{1}{2}\right).$$

The key observation is that each time we move from A_1 to A_2 to A_4 to A_8 to A_{16}, and so forth, the sum of the newly added terms exceeds 1/2. Thus,

$$A_{2^m} > \frac{m+1}{2},$$

and given any $M > 0$, one needs $m > 2M$ to guarantee that $A_{2^m} > \frac{m+1}{2} > \frac{2M+1}{2} > M$ and hence that A_{2^m} and therefore A_n diverge to infinity. □

Example 3.1.1. Consider the series

$$\sum_{k=0}^{\infty} \frac{2}{k+2}.$$

Does this series converge or diverge?

Answer. This is not a particularly famous series as it stands, but sometimes the key step in understanding whether or not a series converges is to realize that it is essentially a series that we already know. Such is the case here. Let $j = k + 2$; then in terms of the partial sums,

$$\sum_{k=0}^{n} \frac{2}{k+2} = \sum_{j=2}^{n+2} \frac{2}{j} = 2\left(\sum_{j=1}^{n+2} \frac{1}{j}\right) - 2.$$

Now taking the limit of these partial sums as $n \to \infty$, one see that the series in this example is 2 less than twice the harmonic series, and since the harmonic series diverges, so must this series.

Remarks.

1. The above example is a good example of how a series can be *reindexed* (the index of the series can be shifted), and a finite number of terms can be added or subtracted to transform a given series into one that is already understood.

2. It is tempting to just write

$$\sum_{k=0}^{\infty} \frac{2}{k+2} = 2\left(\sum_{j=1}^{\infty} \frac{1}{j}\right) - 2$$

from the start, leaving out the use of the definition of series in terms of partial sums. Of course this displayed statement is true, at least in the sense that both series diverge to infinity, but for the moment, we do not have any general theorems on how to manipulate series, so to be careful, we need to base everything we do with series on their definition in terms of sequences of partial sums. While this series manipulation is valid, as we shall see, there are other seemingly innocent manipulations that *are not valid*.

3.2 Geometric series

There is one type of series where not only convergence can be quickly determined, but also the value of the convergent series can be found:

Definition. Geometric series: Given real constants a and r (for *ratio*), any series of the form

$$\sum_{k=0}^{\infty} ar^k := \lim_{n \to \infty} \sum_{k=0}^{n} ar^k \equiv \lim_{n \to \infty} (a + ar + ar^2 + \cdots + ar^n).$$

Remark. Notice that the *ratio* of two consecutive terms in this series is

$$\frac{ar^{k+1}}{ar^k} = r$$

regardless of the value of k; this justifies the choice of the letter r. Also notice that a is the leading term in the series (here, the zeroth term).

Technically the definition of this series only makes sense if the limit of the sequence of partial sums converges, so the next theorem is key to understanding geometric series.

Theorem 3.2.1 (Geometric series). *If $|r| < 1$ (i.e., $-1 < r < 1$), then*

$$\sum_{k=0}^{\infty} ar^k = \frac{a}{1-r}.$$

On the other hand, if $|r| \geq 1$, then the geometric series diverges.

Proof. Using the definition of the series as the limit of the sequence of partial sums, when $|r| < 1$, one can creatively multiply by $1 = (1-r)/(1-r)$ to form a telescoping sum:

$$\sum_{k=0}^{\infty} ar^k = \lim_{n\to\infty}(a + ar + ar^2 + \cdots + ar^n)\frac{1-r}{1-r} = a\lim_{n\to\infty}\frac{1-r^{n+1}}{1-r} = \frac{a}{1-r},$$

because $\lim_{n\to\infty} r^{n+1} = 0$ when $|r| < 1$. On the other hand, if $|r| \geq 1$, this limit and the entire series must diverge. $\square$

Several basic examples are perhaps helpful to understand how to work with geometric series.

Example 3.2.1. Consider the series

$$\sum_{k=1}^{\infty}\left(\frac{1}{7}\right)^k = \frac{1}{7} + \frac{1}{49} + \frac{1}{343} + \cdots.$$

Does this series converge, and if so, to what limit?

Answer. Notice that this series is almost but not quite of the form given in the definition of geometric series. The issue is that this series starts with $k = 1$, not $k = 0$. To take care of this issue, factor out $1/7$ from each term of the series. Then we have

$$\sum_{k=1}^{\infty}\left(\frac{1}{7}\right)^k = \sum_{k=1}^{\infty}\frac{1}{7}\left(\frac{1}{7}\right)^{k-1} = \sum_{n=0}^{\infty}\frac{1}{7}\left(\frac{1}{7}\right)^n,$$

where $n = k - 1$. So this is a geometric series with $a = 1/7$ and $r = 1/7$ too, and by the geometric series theorem,

$$\sum_{k=1}^{\infty}\left(\frac{1}{7}\right)^k = \frac{1}{7} + \frac{1}{49} + \frac{1}{343} + \cdots = \frac{\frac{1}{7}}{1-\frac{1}{7}} = \frac{1}{6}.$$

Notice that in the example above, the key was to realize that one needs to factor out $1/7$ and let $n = k - 1$ to rewrite the series in the exact form that was needed. As we will see throughout the rest of the examples in this section, this is typically the case. One needs to be able to spot how to rewrite the series to get it in the form we need. But notice that we did not reorder the terms (commute the terms) in the series; as we will see later, reordering will change the value of certain series.

Example 3.2.2. Consider the series

$$\sum_{k=1}^{\infty}(-1)^k\left(\frac{2}{e}\right)^k.$$

Does this series converge, and if so, to what limit?

Answer. Notice that the terms in this series can be rewritten so that it is a geometric series with $a = -2/e$ and $-1 < r = -2/e < 1$:

$$\sum_{k=1}^{\infty}(-1)^k\left(\frac{2}{e}\right)^k = \sum_{k=1}^{\infty}\left(\frac{-2}{e}\right)^k = \sum_{n=0}^{\infty}\left(\frac{-2}{e}\right)\left(\frac{-2}{e}\right)^n = \frac{-2}{e}\left(\frac{1}{1+\frac{2}{e}}\right) = \frac{-2}{e+2}.$$

Here again $k = n + 1$.

Example 3.2.3. Finally, in this set of examples, consider

$$\sum_{k=0}^{\infty} 2^k = 1 + 2 + 4 + 8 + \cdots.$$

Does this series converge, and if so, to what value?

Answer. A quick look at this series and its partial sums tells us that this series must diverge. In this case, the wrong approach is to observe that this is a geometric series (which it is) and conclude that since $a = 1$ and $r = 2$,

$$\sum_{k=0}^{\infty} 2^k = 1 + 2 + 4 + 8 + \cdots = \frac{1}{1-2} = -1. \quad \textbf{This is False!}$$

The problem of course is that here $r > 1$, so the convergence portion of the geometric series theorem dose not apply. Indeed, how could a series where every term is positive possibly converge to a negative number? The answer is that it cannot! As we see below, this series also fails the very basic n-th term test: the n-th term here does not converge to zero. If any sense is to be made of this geometric series, it requires a completely different understanding of series and convergence.

Before moving on to series convergence tests, there is one more geometric series to consider—one that came up in the previous chapter.

Example 3.2.4. Show that $0.49999\ldots = 0.4\bar{9} = 0.5$. Thus, all terminating decimals have an alternative numerical representation.

Answer. The key to understanding this equality is to view it as a geometric series:

$$0.49999\ldots = \frac{4}{10} + \frac{9}{10^2} + \frac{9}{10^3} + \frac{9}{10^4} + \frac{9}{10^5} + \cdots = \frac{4}{10} + 9\sum_{k=2}^{\infty}\frac{1}{10^k} = \frac{4}{10} + \left(\frac{9/100}{1-1/10}\right) = 0.5.$$

3.3 Convergence tests

This rather long section discusses the many tests that are used to determine whether various series converge or diverge. Some of these tests depend on a series having a specific form (e. g., alternating series), and others require us to carry out some action (integration, comparison, etc.).

3.3.1 Test 1: *n*-th term test

The next result is very basic, but must be understood carefully.

Theorem 3.3.1 (*n*-th term test). *If the series*

$$\sum_{n=1}^{\infty} a_n$$

converges, then the sequence of n-th terms $\{a_n\}$ must converge to 0.

Proof. The proof of this result is again based on the partial sums. First notice that

$$A_n = \sum_{k=1}^{n} a_k = a_n + \sum_{k=1}^{n-1} a_k = a_n + A_{n-1},$$

that is, the *n*-th term can be split out of the sum. Now suppose the series converges to $L \in \mathbb{R}$. By definition, then, given any $\epsilon > 0$, $\exists N \in \mathbb{Z}^+$ such that $|A_n - L| < \epsilon/2$ whenever $n > N + 1$. The choices of $\epsilon/2$ and $N + 1$ are again by design to make the end of the proof work out nicely. Other than these choices, however, this is just the standard definition of the limit of a sequence. Now by the triangle inequality, given any $\epsilon > 0$, $\exists N \in \mathbb{Z}^+$ such that

$$|a_n| = |A_n - A_{n-1}| = \left|(A_n - L) + (L - A_{n-1})\right| \le |A_n - L| + |A_{n-1} - L| < \frac{\epsilon}{2} + \frac{\epsilon}{2} = \epsilon$$

whenever $n > N + 1$ (so that both n and $n - 1$ exceed N). Thus, a_n becomes arbitrarily small as n increases, or in symbols, $a_n \to 0$ as $n \to \infty$. □

Remark. Notice that the *n*-th term test works in one direction only: the series converging implies that the *n*-th term goes to zero, but the *n*-th term going to zero does not imply that the series converges. In particular, the *n*-th term in the harmonic series $(1/n)$ goes to zero, but we just showed that the harmonic series diverges. The *n*-th term test can never be used to show convergence.

Corollary 3.3.1. *If $a_n \not\to 0$, then the series diverges.*

Proof. Notice that this corollary is just the contrapositive of the *n*-th term test, so it is automatically true. □

3.3.2 Test 2: alternating series test

After geometric series, probably the most recognizable type of series is the alternating series where the signs of the terms alternate between positive and negative.

Definition. **Alternating series**: Given a sequence $\{a_k\}$ with $a_k > 0 \; \forall \, k \in \mathbb{N}$, any series of the form

$$\sum_{k=0}^{\infty} (-1)^k a_k := \lim_{n \to \infty} \sum_{k=0}^{n} (-1)^k a_k \equiv \lim_{n \to \infty} (a_0 - a_1 + a_2 - a_3 + \cdots + (-1)^n a_n).$$

Remark. Notice that unlike in the general notation for series, here the $(-1)^k$ is separate from a_k. Since all the a_k are positive, the terms in this series alternate between positive and negative.

Again, the definition above makes sense exactly when the series and the limit converge.

Theorem 3.3.2 (Leibniz[1] alternating series test). *Suppose that the sequence $\{a_k\}$ is **decreasing**, that is, $a_k \geq a_{k+1}$, and that $a_k \to 0$. Then the alternating series*

$$\sum_{k=0}^{\infty} (-1)^k a_k$$

converges to a real number L, with $A_{2n+1} \leq L \leq A_{2n} \; \forall \, n \in \mathbb{Z}^+$.

Proof. The key observation is that for a given partial sum A_n, the next partial sum A_{n+1} is obtained by either adding or subtracting a_{n+1} to/from A_n (depending on whether n is odd or even), and a_{n+1} is smaller than a_n, so its addition or subtraction cannot get us back to A_{n-1}. Thus, $A_0 = a_0$ is the largest partial sum, $A_1 = A_0 - a_1$ must be nonnegative, $A_2 = A_1 + a_2$ cannot exceed A_0, and so forth, back and forth. This arrangement is shown in Figure 3.1. Now what is needed is a completely rigorous derivation that fully implements the above intuition.

Consider the odd and even partial sums A_{2n+1} and A_{2n} for $n \in \mathbb{N}$. Notice that

$$A_{2n+1} = A_{2n-1} + a_{2n} - a_{2n+1} \quad \Longleftrightarrow \quad A_{2n+1} - A_{2n-1} = a_{2n} - a_{2n+1} > 0$$

while

$$A_{2n+2} = A_{2n} - a_{2n+1} + a_{2n+2} \quad \Longleftrightarrow \quad A_{2n+2} - A_{2n} = a_{2n+2} - a_{2n+1} < 0,$$

---

[1] Gottfried Leibniz (1646–1716) was a German philosopher and mathematician generally credited (along with Isaac Newton) with discovering calculus.

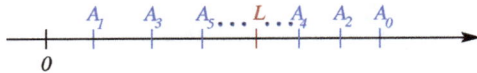

Figure 3.1: For an alternating series with a_n decreasing to zero, the even partial sums form a decreasing sequence that heads for L coming down from above, while the odd partial sums form an increasing sequence that heads for L coming up from below.

because the sequence $\{a_k\}$ is itself decreasing. Thus, the odd partial sums form an increasing sequence, while the even partial sums form a decreasing sequence. Moreover, the odd partial sums are bounded above by $A_0 = a_0$, while the even partial sums are bounded below by $A_1 = a_0 - a_1$ (again these bounds are because $\{a_k\}$ is itself decreasing). So by the monotone convergence theorem, the odd partial sums must converge to some real number L_-, while the even partial sums must converge to a real number L_+.

The final detail is to show that $L_- = L_+$. This is accomplished by using the relationship

$$A_{2n+1} = A_{2n} - a_{2n+1} \quad \Longleftrightarrow \quad A_{2n} - A_{2n+1} = a_{2n+1}$$

and noting that $a_k \to 0$:

$$L_+ - L_- = \lim_{n\to\infty} A_{2n} - \lim_{n\to\infty} A_{2n+1} = \lim_{n\to\infty} (A_{2n} - A_{2n+1}) = \lim_{n\to\infty} a_{2n+1} = 0.$$

Hence, *all* the partial sums must converge to the same real number $L := L_- = L_+$. That $A_{2n+1} \leq L \leq A_{2n}$ comes from the calculation above since the odd partial sums increase from below to L, while the even partial sums decrease from above. □

Example 3.3.1. Is the series

$$\sum_{n=0}^{\infty} \frac{(-1)^n}{(n+1)^{3/2}} = \sum_{n=0}^{\infty} \frac{(-1)^n \sqrt{n+1}}{(n+1)^2} = 1 - \frac{\sqrt{2}}{4} + \frac{\sqrt{3}}{9} - \frac{1}{8} + \cdots$$

an alternating series? Does it converge, and if so, what can be said about the value of the limit?

Answer. Notice that the series exactly satisfies the form of an alternating series with

$$a_n = (n+1)^{-3/2}$$

decreasing to zero as $n \to \infty$, so this series must converge. While the alternating series test does not say what the limiting value L is, it does say that

$$A_5 = 1 - \frac{\sqrt{2}}{4} + \frac{\sqrt{3}}{9} - \frac{1}{8} + \frac{\sqrt{5}}{25} - \frac{\sqrt{6}}{36} \leq L \leq 1 - \frac{\sqrt{2}}{4} + \frac{\sqrt{3}}{9} - \frac{1}{8} + \frac{\sqrt{5}}{25} = A_4$$

or $0.70 < L < 0.81$. One could obtain tighter bounds by choosing $n > 2$. For example, for $n = 5$, one finds that $A_{11} \approx 0.7539 < L < 0.7779 \approx A_{10}$.

Example 3.3.2. Does the series

$$\sum_{k=1}^{\infty} \frac{(-1)^{k-1}}{k} = 1 - \frac{1}{2} + \frac{1}{3} - \frac{1}{4} + \cdots$$

converge or diverge? If it converges, to what value?

Answer. As written, this series does not quite fit the form given above, but notice that this is only a small detail because this series can be rewritten as

$$\sum_{n=0}^{\infty} \frac{(-1)^n}{n+1} = 1 - \frac{1}{2} + \frac{1}{3} - \frac{1}{4} + \cdots$$

simply by reindexing this series using $k = n + 1$. One should keep in mind that a series may need to be reindexed in order to recognize its form as an alternating series.

This series is often called the **alternating harmonic series**. Recall that the harmonic series diverges; the alternating series test, however, guarantees that the alternating harmonic series converges. Notice that the terms in the series do alternate in sign, and here

$$a_n = \frac{1}{n+1},$$

which is strictly decreasing to 0 as n increases. But again the alternating series test does not give us the value of this series. We will have to wait until our discussion of Taylor series to understand why, but the value of the alternating series is known:

$$\sum_{k=1}^{\infty} \frac{(-1)^{k-1}}{k} = \ln(2).$$

The previous example leads to a question: Which series converge because of cancelation between their positive and negative terms, but would diverge if signs of the negative terms are all flipped, making all terms positive? This cancelation is the reason why the alternating harmonic series converges even though the harmonic series diverges. The next definition and theorem deal with this issue.

Definition. Suppose that the series

$$\sum_{k=1}^{\infty} a_k$$

converges. If the series

$$\sum_{k=1}^{\infty} |a_k|$$

also converges, then the original series is said to **converge absolutely**, or to be **absolutely convergent**. Alternatively, if the latter series diverges, then the original series is said to **converge conditionally**, or to be **conditionally convergent**. That is, the original series converges only because of cancelation.

Our previous example, Example 3.3.2, shows that the alternating harmonic series is conditionally convergent, while the geometric series in Example 3.2.2 is absolutely convergent.

One might wonder if there are any series where

$$\sum_{k=1}^{\infty} a_k$$

diverges, even though

$$\sum_{k=1}^{\infty} |a_k|$$

converges. This never happens.

Theorem 3.3.3. *We have the following relationships:*

$$\sum_{k=1}^{\infty} |a_k| \text{ converges} \quad \Longrightarrow \quad \sum_{k=1}^{\infty} a_k \text{ converges},$$

$$\sum_{k=1}^{\infty} a_k \text{ diverges} \quad \Longrightarrow \quad \sum_{k=1}^{\infty} |a_k| \text{ diverges}.$$

Proof. The second statement is just the contrapositive of the first, so it is only necessary for us to prove the first. The key to proving the first statement is to break the terms of this series into their positive and negative parts: For each $k \in \mathbb{Z}^+$, let

$$p_k := \begin{cases} a_k & \text{if } a_k \geq 0 \\ 0 & \text{if } a_k \leq 0 \end{cases} \quad \text{and} \quad n_k := \begin{cases} 0 & \text{if } a_k \geq 0 \\ -a_k & \text{if } a_k \leq 0. \end{cases}$$

Notice that both p_k and n_k are themselves nonnegative: $a_k = p_k - n_k$ and $|a_k| = p_k + n_k$, where $p_k, n_k \geq 0 \, \forall k \in \mathbb{Z}^+$. This implies that $0 \leq p_k \leq |a_k|$ and $0 \leq n_k \leq |a_k|$, and hence for the partial sums,

$$\sum_{k=1}^{n} p_k \leq \sum_{k=1}^{n} |a_k| \quad \text{and} \quad \sum_{k=1}^{n} n_k \leq \sum_{k=1}^{n} |a_k|.$$

Now all of the terms in each of these sums is positive, and the series in a_k is absolutely convergent, so $\exists A \in \mathbb{R}$ such that

$$\sum_{k=1}^{n} |a_k| \nearrow A = \sum_{k=1}^{\infty} |a_k|.$$

Thus both

$$\sum_{k=1}^{n} p_k \quad \text{and} \quad \sum_{k=1}^{n} n_k$$

are increasing sequences bounded above by A, and therefore each must converge to a real number. Since these two sequences are the partial sums of the positive and negative series, both

$$\sum_{k=1}^{\infty} p_k \quad \text{and} \quad \sum_{k=1}^{\infty} n_k$$

must converge.

Finally, recalling that the limit of a difference is the difference of the limits, one finds

$$\sum_{k=1}^{\infty} a_k = \lim_{n \to \infty} \sum_{k=1}^{n} a_k = \lim_{n \to \infty} \sum_{k=1}^{n} (p_k - n_k) = \lim_{n \to \infty} \left(\sum_{k=1}^{n} p_k - \sum_{k=1}^{n} n_k \right)$$

$$= \left(\lim_{n \to \infty} \sum_{k=1}^{n} p_k \right) - \left(\lim_{n \to \infty} \sum_{k=1}^{n} n_k \right) = \sum_{k=1}^{\infty} p_k - \sum_{k=1}^{\infty} n_k,$$

since both of these final two series converge. □

The previous theorem inspires a well-known (if corny) joke: *Question*: Does an absolutely convergent series converge? *Answer*: Absolutely!

The previous proof uses an upper bound on an increasing sequence to prove convergence. This approach not only yields the previous result showing that absolute convergence implies convergence, it also leads to our next convergence tests: the comparison tests.

3.3.3 Test 3: comparison tests

The next test is really a pair of tests: the comparison tests. The basic or direct comparison test determines the convergence or divergence of a given series by direct term-by-term comparison with a series whose convergence or divergence is known. This test is simple and direct, but sometimes hard to use because it requires a term-by-term comparison. To avoid this issue, the second comparison test, the limit comparison test, requires only the consideration of a certain limit described below.

Theorem 3.3.4 (Direct comparison test). *For each $k \in \mathbb{Z}^+$, suppose that $a_k \in \mathbb{R}$ and that $b_k, M \geq 0$. Finally, suppose that there exist $N \in \mathbb{Z}^+$ such that for each $k > N$,*

$$|a_k| \le Mb_k.$$

Then

$$\sum_{k=1}^{\infty} b_k \text{ converges} \quad \Longrightarrow \quad \sum_{k=1}^{\infty} |a_k| \text{ converges} \quad \Longrightarrow \quad \sum_{k=1}^{\infty} a_k \text{ converges,}$$

and equivalently

$$\sum_{k=1}^{\infty} a_k \text{ diverges} \quad \Longrightarrow \quad \sum_{k=1}^{\infty} |a_k| \text{ diverges} \quad \Longrightarrow \quad \sum_{k=1}^{\infty} b_k \text{ diverges.}$$

Proof. Notice that the second statement is again the contrapositive of the first. Also, the second implication of the first statement is just that absolute convergence implies convergence (Theorem 3.3.3). Thus only the first implication of the first statement requires proof.

The key assumption is the term-by-term bound for the terms of one sequence by some multiple of the terms of the other. So in terms of partial sums,

$$|a_k| \le Mb_k \quad \Longrightarrow \quad \sum_{k=N+1}^{n} |a_k| \le M \sum_{k=N+1}^{n} b_k$$

for all $n > N$. Because the series in b_k is assumed to converge, so does the series in b_k that begins at $N+1$ (the latter is just the entire series minus the finite sum of the first N terms). Let

$$B := \sum_{k=N+1}^{\infty} b_k.$$

Then $\forall\, n > N$,

$$\sum_{k=N+1}^{n} |a_k| \le M \sum_{k=N+1}^{n} b_k \le M \sum_{k=N+1}^{\infty} b_k = MB \quad \Longrightarrow \quad \sum_{k=1}^{n} |a_k| \le MB + \sum_{k=1}^{N} |a_k|$$

(the first N terms are simply added to the original bound to yield a bound for the partial sums corresponding to the entire series in $|a_k|$). Since the partial sums for this entire series form an increasing sequence that is bounded above, both these partial sums and the corresponding series must converge. □

Example 3.3.3. Does the series

$$\sum_{k=1}^{\infty} \frac{1}{\sqrt{k}}$$

converge or diverge?

Answer. Determining the convergence or divergence of this series is probably most easily done using the direct comparison test. Notice that for all $k \in \mathbb{Z}^+$

$$\sqrt{k} \le k \quad \Longrightarrow \quad \frac{1}{k} \le \frac{1}{\sqrt{k}},$$

which means that by the comparison test, the divergence of the harmonic series implies the divergence of the series at hand in this example.

The comparison test works well when there is a direct term-by-term comparison; it works much less well when this sort of comparison is difficult or impossible. Consider the next example.

Example 3.3.4. Does the series

$$\sum_{k=1}^{\infty} \frac{1}{(\sqrt{k}+1)^2}$$

converge or diverge?

Answer. Because of the square root and the square in the denominator, one might expect that this series would diverge just as the harmonic series does. Unfortunately, for all $k \in \mathbb{Z}^+$

$$k \le (\sqrt{k}+1)^2 \quad \Longrightarrow \quad \frac{1}{(\sqrt{k}+1)^2} \le \frac{1}{k},$$

which is exactly the wrong inequality to apply the direct comparison test. Luckily, there is a version of the comparison test that does not require a strict term-by-term comparison. Rather, it requires that a certain limit be computed.

Theorem 3.3.5 (Limit comparison test). *Suppose that $a_k \in \mathbb{R}$ and $b_k > 0\ \forall k \in \mathbb{Z}^+$ and that*

$$\lim_{k \to \infty} \frac{|a_k|}{b_k}$$

exists and is finite. Then

$$\sum_{k=1}^{\infty} b_k\ converges \quad \Longrightarrow \quad \sum_{k=1}^{\infty} |a_k|\ converges \quad \Longrightarrow \quad \sum_{k=1}^{\infty} a_k\ converges,$$

and equivalently

$$\sum_{k=1}^{\infty} a_k\ diverges \quad \Longrightarrow \quad \sum_{k=1}^{\infty} |a_k|\ diverges \quad \Longrightarrow \quad \sum_{k=1}^{\infty} b_k\ diverges.$$

We return to the example above.

Example 3.3.4. Does the series

$$\sum_{k=1}^{\infty} \frac{1}{(\sqrt{k}+1)^2}$$

converge or diverge?

Answer. Because the harmonic series diverges, the limit that must be considered to apply the limit comparison test is

$$\lim_{k\to\infty} \frac{1/k}{1/(\sqrt{k}+1)^2} = \lim_{k\to\infty} \frac{(\sqrt{k}+1)^2}{k} = \lim_{k\to\infty}\left(1+\frac{1}{\sqrt{k}}\right)^2 = 1.$$

Because this limit is finite, the limit comparison test ensures that the series in this example must diverge, just as the harmonic series diverges.

Proof of limit comparison test. As in the comparison test, the second statement is the contrapositive of the first, and the second part of the first statement is always true by Theorem 3.3.3. Let $L \in \mathbb{R}$ be the value of the limit in the hypothesis. Since this limit exists, for $\epsilon = 1$, there exists $N \in \mathbb{Z}^+$ such that

$$\left|\frac{|a_k|}{b_k} - L\right| < 1 \quad\Longrightarrow\quad L - 1 < \frac{|a_k|}{b_k} < L + 1 \quad\Longrightarrow\quad |a_k| < (L+1)b_k$$

for all $k > N$. The final bound yields the hypothesis of the direct comparison test, meaning that the direct comparison test now implies the validity of the limit comparison test. □

3.3.4 Test 4: integral test

The next test determines the convergence or divergence of a series by determining the convergence or divergence of an appropriate improper integral. This test in turn is the key tool for determining the convergence or divergence of an important type of series called *p*-series. The problem with this test is that one must be able to compute an improper integral. Our presentation here begins by defining *p*-series and then states the theorem (the integral test). We then digress to study certain improper integrals, before returning (1) to use the integral test to determine the convergence of *p*-series, (2) to consider a set of examples and (3) to finally prove the integral test.

At first glance, *p*-series may look like geometric series, but there is a very important difference.

Definition. *p*-**Series**: Given a real constant p, any series of the form

$$\sum_{k=1}^{\infty} \frac{1}{k^p} := \lim_{n\to\infty} \sum_{k=1}^{n} \frac{1}{k^p} \equiv \lim_{n\to\infty}\left(1 + \frac{1}{2^p} + \frac{1}{3^p} + \cdots + \frac{1}{n^p}\right).$$

Remark. The key difference between a geometric series and a p-series is that in a geometric series, the index is the power, while in a p-series, the power is the constant p.

Determining whether or not p-series converge requires the integral test; it is first stated, then discussed and used to determine the convergence of p-series and finally proven.

Theorem 3.3.6 (Maclaurin–Cauchy integral test). *Suppose that $f : [1, +\infty) \to [0, +\infty)$ is a continuous, decreasing function. So $f(x) \geq 0 \ \forall x \geq 1$. Suppose also that $a_n = f(n)$. Then the series*

$$\sum_{n=1}^{\infty} a_n$$

and the improper integral

$$\int_1^{\infty} f(x)\, dx$$

both converge or diverge together. That is, if either converges, they both converge, and if either diverges, they both diverge.

Students reading the above theorem will notice that there is a term in its statement that is likely new: *improper integral*. Up until this point, all of the integrals that most students have likely encountered have been standard (proper) Riemann integrals. Now we need to define the kind of improper integral that appears in the integral test. For the moment, only continuous integrands are considered.

Definition. Suppose that $a \in \mathbb{R}$ is fixed and that for all $b > a$, the function f is continuous on $[a, b]$. An **improper Riemann integral of the first kind** is an extension of the standard (definite) Riemann integral to a semi-infinite domain:

$$\int_a^{\infty} f(x)\, dx := \lim_{b \to \infty} \int_a^b f(x)\, dx,$$

provided that this limit exists *as a finite real number.*

Improper integrals will be dealt with much more thoroughly in Chapter 5; for the moment, we will just consider two basic examples that are sufficient for us to study p-series and the integral test.

Example 3.3.5. Consider the integral

$$\int_1^{\infty} \frac{dx}{x^2}.$$

Does this integral converge, and if so, to what value?

Answer. This is an improper integral because the upper limit of integration is infinity; it must be defined and computed as a limit:

$$\int_1^\infty \frac{dx}{x^2} := \lim_{b\to\infty} \int_1^b x^{-2}\, dx = \lim_{b\to\infty} \left(\frac{x^{-1}}{-1}\Big|_1^b\right) = \lim_{b\to\infty}\left(1 - \frac{1}{b}\right) = 1 - 0 = 1.$$

Notice that the integral inside the limit is just a standard Riemann integral, so it can be computed using standard integral calculus as was done here.

Example 3.3.6. Consider the integral

$$\int_1^\infty \frac{dx}{x}.$$

Does this integral converge, and if so, to what value?

Answer. Again, this is an improper integral, so it must also be defined as a limit:

$$\int_1^\infty \frac{1}{x}\, dx := \lim_{b\to\infty} \int_1^b x^{-1}\, dx = \lim_{b\to\infty} \ln b.$$

Those who are not familiar with the integral inside this limit might be tempted to apply the power rule for integration. But the power here is the one that is not allowed by that rule. In fact, this integral is often taken as the definition of the natural logarithm function (ln) as is done here. The key property of the logarithm that is needed here is that it is unbounded as its argument grows. So the above limit does not exist as a finite real number, and hence the improper integral in this example also does not converge. The natural logarithm function will be discussed more thoroughly in Section 5.1 below.

The previous two examples were not chosen randomly. According to the integral test, the key integral for us to consider to determine the convergence or divergence of a p-series is

$$\int_1^\infty \frac{dx}{x^p} := \lim_{b\to\infty} \int_1^b x^{-p}\, dx.$$

Based on our previous examples, we already know the value of this limit for $p = 2$ and $p = 1$. Except when $p = 1$, the integral inside the limit above can be found using the power rule:

$$\int_1^\infty \frac{dx}{x^p} := \lim_{b\to\infty} \int_1^b x^{-p}\, dx = \lim_{b\to\infty}\left(\frac{x^{1-p}}{1-p}\Big|_1^b\right) = \frac{1}{p-1} - \lim_{b\to\infty} \frac{b^{1-p}}{p-1}.$$

Not surprisingly, the convergence or divergence of this final limit depends on the value of p: When $p > 1$, the final limit above is zero, the improper integral converges, and by the integral test, so does the p-series. When $p < 1$, the final limit diverges, the improper integral diverges, and by the integral test, the p-series diverges.

Remark. For $p > 1$, the integral can be evaluated exactly, but this does not tell us the value of the corresponding p-series. The integral test only tells us that the series and the integral both converge or diverge together. When they both converge, their values are not likely to be the same.

Our work above implies the following theorem.

Theorem 3.3.7 (p-Series). *A p-series converges when $p > 1$ and diverges when $p \le 1$.*

There is one thing left for us to discuss: Why is the integral test valid? As will become clear in the proof and the diagram in Figure 3.2, the integral test is really a kind of comparison test. Since an integral represents the area under the curve $y = f(x)$, this area can be bounded both above and below by versions of the series.

Proof of Maclaurin–Cauchy integral test. The proof of this result is again based on the partial sums and is depicted graphically in Figure 3.2: First suppose that the improper integral converges, meaning that

$$\int_1^\infty f(x)\, dx = \lim_{n\to\infty} \int_1^n f(x)\, dx$$

is a finite real number. Notice that the red rectangular boxes in Figure 3.2 lie entirely below the curve $y = f(x)$, implying that

$$A_n = a_1 + \sum_{k=2}^n a_k \le a_1 + \int_1^n f(x)\, dx,$$

that is, if the first term in the partial sum is set aside, the proper integrals up to n bound A_n, the partial sums up to n. Since these proper integrals converge to the improper integral and a_n are nonnegative, the partial sums A_n form an increasing sequence that is bounded above, so it must converge by the monotone convergence theorem.

Now suppose the series converges to $L \in \mathbb{R}$, meaning that both $A_n \nearrow L$ and $A_{n-1} \nearrow L$. The partial sum A_{n-1} is represented in Figure 3.2 by the green rectangular boxes (extending down behind the red boxes to the x-axis). So $\forall\, n \in \mathbb{Z}^+$,

$$\int_1^n f(x)\, dx \le A_{n-1} \le L,$$

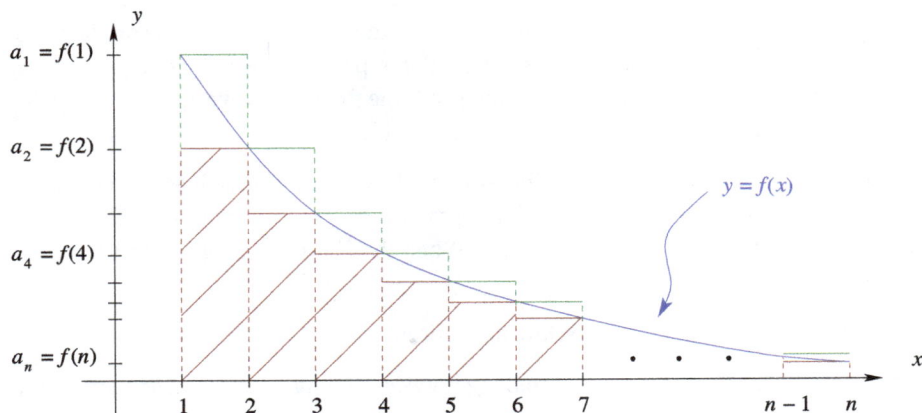

Figure 3.2: Bounds for the integral test. The area in/under the green rectangular boxes represents the partial sum A_{n-1}. The area under the blue curve represents the integral from 1 to n. The area in/under the red rectangular boxes represents the partial sum A_n, except that a_1 is excluded, so this area is $A_n - a_1$.

implying that the proper integrals from 1 to n form an increasing sequence that is bounded above. Hence the limit of this sequence (the improper integral) must exist as a finite number.

Combining these two bounding results, one finds that an improper integral and the series must always converge or diverge together. □

Although the most important use of the integral test is in determining when p-series converge, there are some other series where the integral test can be helpful. The important thing in using the integral test in this way is that the appropriate integral must be evaluated, or at least bounded.

Example 3.3.7. Does the series

$$\sum_{n=2}^{\infty} \frac{1}{n(\ln n)^2}$$

converge or diverge?

Answer. This series clearly is not of the form of a geometric series, an alternating series or a p-series, so whether it converges is not apparent. That it does converge is most easily seen by applying the integral test. Consider

$$\int_2^{\infty} \frac{dx}{x(\ln x)^2}.$$

This may seem to be an unusual integral, but it is relatively easy to evaluate using the substitution $u = \ln x$. Then because $du = dx/x$, we find

$$\int\limits_{2}^{\infty} \frac{dx}{x(\ln x)^2} = \int\limits_{\ln 2}^{\infty} \frac{du}{u^2} = -\left(\frac{1}{u}\Big|_{\ln 2}^{\infty}\right) = \frac{1}{\ln 2}.$$

Again the value of this integral is not the value of the series; the convergence of this improper integral simply implies the convergence of the series. Also, one might notice here that the improper integral was evaluated without explicitly writing it as a limit. This shorthand notation, treating infinity as a finite limit of integration, is often convenient and widely used, especially outside mathematics. The important thing is to keep in mind that improper integrals at times must be explicitly written as limits when this shorthand notation does not seem to work properly.

3.3.5 Test 5: ratio test

The next test to be considered, the ratio test, is especially useful when the terms in a series have powers or factorials of the index k. This is important because perhaps the most important function series, the power series, often contains both these powers and factorials.

Theorem 3.3.8 (d'Alembert[2]–Cauchy ratio test). *Suppose that $a_k \neq 0 \; \forall \, k \in \mathbb{Z}^+$ and that*

$$r := \lim_{k \to \infty} \left| \frac{a_{k+1}}{a_k} \right|$$

exists as a finite number or is $+\infty$. Then

- *If $r < 1$, then $\sum_{k=1}^{\infty} |a_k|$ converges, and hence $\sum_{k=1}^{\infty} a_k$ also converges.*
- *If $r > 1$, then $\sum_{k=1}^{\infty} a_k$ diverges, and hence $\sum_{k=1}^{\infty} |a_k|$ also diverges.*
- *If $r = 1$, then the test is inclusive (i. e., the test fails).*

Proof. If $r < 1$, fix $\rho \in (r, 1)$ (for example, let $\rho = (1 + r)/2$ be the midpoint), and let $\epsilon := \rho - r$ (the distance between ρ and r). Then since $|a_{k+1}/a_k|$ converges to r, $\exists \, N \in \mathbb{Z}^+$ such that $\forall \, k > N - 1$,

$$\left| \left| \frac{a_{k+1}}{a_k} \right| - r \right| < \epsilon.$$

See the diagram in Figure 3.3. Hence

$$\left| \frac{a_{k+1}}{a_k} \right| < \rho \quad \Longrightarrow \quad |a_{k+1}| < \rho \, |a_k|.$$

---

2 Jean le Rond d'Alembert (1717–1783) was a French mathematician, physicist and music theorist known for his solution to the basic wave equation.

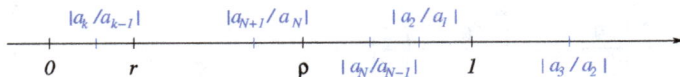

Figure 3.3: Sequence ratios converging to r. There is no particular control on the ratios for the first few elements in the sequence, but for $k > N$, each ratio is less than ρ. Where should ϵ be placed on this diagram? Notice that this diagram is the same as the one in Figure 1.4, but the proof that it illustrates here is somewhat different.

So

$$|a_{N+1}| < \rho\,|a_N|$$

and

$$|a_{N+2}| < \rho\,|a_{N+1}| < \rho^2\,|a_N|,$$

and in general

$$|a_{N+j}| < \rho^j\,|a_N|.$$

Now let $k = N + j$; then $\forall k > N$,

$$|a_k| < \rho^{(k-N)}\,|a_N| = M\,\rho^k,$$

where $M := |a_N|/\rho^N$. This is of course exactly the requirement to apply the direct comparison test; the series in $|a_k|$ thus converges by comparison to the geometric series with $\rho < 1$.

For $r > 1$, the proof of divergence is discussed in Exercise B.3.9. For $r = 1$, notice that the harmonic series diverges, while the p-series with $p = 2$ converges, even though both have the same ratio $r = 1$. □

Example 3.3.8. Does the series

$$\sum_{k=0}^{\infty} \frac{5^k}{k!}$$

converge or diverge?

Answer. Since k appears both as a power and in a factorial, this series is begging to have the ratio test applied to it. Applying the ratio test, one must consider the limit

$$\lim_{k\to\infty} \frac{a_{k+1}}{a_k} = \lim_{k\to\infty} \frac{\frac{5^{k+1}}{(k+1)!}}{\frac{5^k}{k!}} = \lim_{k\to\infty} \frac{5^{k+1}}{5^k}\frac{k!}{(k+1)!} = \lim_{k\to\infty} \frac{5}{k+1} = 0.$$

Since the value of this limit is less than 1, this series must converge.

The exact value of this series is not given by the ratio test (or any other test). In the next chapter, the value will be found as a Taylor series; the value is e^5.

3.3.6 Test 6: root test

There is one final convergence test that is commonly studied: the root test. Strictly speaking, this test is not needed for our discussion on sequences and series of functions, but still there are situations where it can be useful. It is also only a small variation on the ratio test that was just discussed. In general, the root test is most useful when an expression (perhaps a complicated expression) is raised to the k-th power, this power then being canceled by the k-th root.

Theorem 3.3.9 (Cauchy root test). *Suppose that*

$$r := \lim_{k \to \infty} \sqrt[k]{|a_k|}$$

exists as a finite number or is $+\infty$. *Then*
- *If $r < 1$, then $\sum_{k=1}^{\infty} |a_k|$ converges, and hence $\sum_{k=1}^{\infty} a_k$ also converges.*
- *If $r > 1$, then $\sum_{k=1}^{\infty} a_k$ diverges, and hence $\sum_{k=1}^{\infty} |a_k|$ also diverges.*
- *If $r = 1$, then the test is inclusive (i. e., the test fails).*

Proof. For $r < 1$, again fix $\rho \in (r, 1)$ (again, the midpoint $\rho := (1 + r)/2$ is a good choice) and take $\epsilon := \rho - r$. Now since $\sqrt[k]{|a_k|}$ converges to r, $\exists N \in \mathbb{Z}^+$ such that $\forall k > N - 1$,

$$\left| \sqrt[k]{|a_k|} - r \right| < \epsilon.$$

See the diagram in Figure 3.4. Hence

$$\sqrt[k]{|a_k|} < \rho \quad \Longleftrightarrow \quad |a_k| < \rho^k.$$

So the series in $|a_k|$ converges by comparison with the geometric series whose ratio is ρ.

For $r > 1$, a variation on the proof for $r < 1$ shows that the series diverges; see Exercise B.3.10. For $r = 1$, notice that the harmonic series diverges, while the p-series with $p = 2$ converges, even though both have the same root limit $r = 1$. □

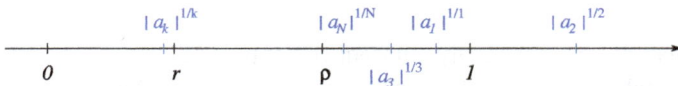

Figure 3.4: Sequence roots converging to r. There is no particular control on the roots for the first few elements in the sequence, but for $k > N$, each root is less than ρ. Again, where should ϵ be placed on this diagram?

Example 3.3.9. Does the series

$$\sum_{m=1}^{\infty} \frac{1}{m^m}$$

converge or diverge?

Answer. Since m appears both as a base and as a power, one might suspect that the root test could be used. The root test requires us to consider the limit

$$\lim_{m\to\infty} \sqrt[m]{|a_m|} = \lim_{m\to\infty} \frac{1}{\sqrt[m]{m^m}} = \lim_{m\to\infty} \frac{1}{m} = 0.$$

Hence, by the root test, this series converges.

Notice that this convergence question could have been answered by a number of tests. For example, one might use the ratio test or one might compare this series to a p-series with $p = 2$.

At this point, the reader may be pleased to know that the root test is the final convergence test to be considered here. What comes next is a discussion of why care must be taken when reordering (commuting) or regrouping (reassociating) terms in a series, at least when some terms are positive and some are negative. The next chapter will apply what we have learned so far about sequences and series of real numbers to sequences and series of real-valued functions. As mentioned at the beginning of this book, the reason for spending all of this time developing sequences and series of functions is to allow us to apply calculus to such functions.

3.4 Regrouping and reordering series

Consider the following (**completely false**) statement:

$$
\begin{aligned}
0 &= 0 + 0 + 0 + \cdots \\
&= (1 - 1) + (1 - 1) + (1 - 1) + \cdots \\
&= 1 - 1 + 1 - 1 + 1 - 1 + \cdots \\
&= 1 + (-1 + 1) + (-1 + 1) + (-1 + 1) \ldots \\
&= 1 + 0 + 0 + 0 + \cdots \\
&= 1.
\end{aligned}
$$

What went wrong? Where did this seemingly innocent regrouping fail? To see the answer to these questions, look at the series on the right, and write each as the limit of its sequence of partial sums:

$$0 + 0 + 0 + \cdots = \sum_{k=1}^{\infty} 0 \qquad \Longrightarrow \qquad S_n = 0,$$

$$(1-1) + (1-1) + (1-1) + \cdots = \sum_{k=1}^{\infty} (1-1) \qquad \Longrightarrow \qquad S_n = 0,$$

$$1 - 1 + 1 - 1 + 1 - 1 + \cdots = \sum_{k=1}^{\infty} (-1)^{k+1} \qquad \Longrightarrow \qquad S_n = \begin{cases} 1 & n \text{ is even} \\ 0 & n \text{ is odd,} \end{cases}$$

$$1 + (1-1) + (1-1) + (1-1) + \cdots = 1 + \sum_{k=1}^{\infty} (1-1) \qquad \Longrightarrow \qquad S_n = 1,$$

$$1 + 0 + 0 + 0 + \cdots = 1 + \sum_{k=1}^{\infty} 0 \qquad \Longrightarrow \qquad S_n = 1.$$

All of this makes clear that the first two series are equal to each other and to zero because all the partial sums are zero. The last two series are also equal to each other, but now both are equal to one because all the partial sums are one. The series in the middle, on the other hand, has partial sums that alternate between zero and one, so this series diverges. The problem is that the regrouping (reassociation) of terms in steps 2 and 4 is not valid because they change the partial sums, thereby changing the outcome of the series. Of course, even this depends on viewing each (1−1) as its own term in the second and fourth series. Still, the statement 0 = 1 must be false and is sometimes called the fundamental contradiction in mathematics. When one reaches this conclusion, all hope is lost.

At this point, one might suspect that regrouping or reordering terms in a series is never possible. Things are not quite this limited. The key to determining when regrouping or reordering of terms is possible is whether a series is absolutely or conditionally convergent; the main result is the Riemann–Dirichlet[3] series theorem.

Theorem 3.4.1 (Riemann–Dirichlet series theorem). *Suppose that*

$$\sum_{k=0}^{\infty} a_k$$

is a conditionally convergent series and that $x \in \mathbb{R}$ is any arbitrarily chosen real number. Then there exists a permutation μ of the terms in the series so that

$$\sum_{k=0}^{\infty} a_{\mu(k)} = x.$$

In addition, there are other permutations so that the permuted series either diverges to $+\infty$ or $-\infty$ or oscillates without approaching any value, finite or infinite. On the other hand, if the series converges absolutely, then any permutation of the terms of the series converges to the same limit.

---

3 Gustav Lejeune Dirichlet (1805–1859) was a German mathematician known for his work on analysis, differential equations and the theory of functions.

Proof. First suppose that the series converges absolutely. As in our proof that absolute convergence implies convergence (Theorem 3.3.3), for each $k \in \mathbb{Z}^+$, define the positive and negative terms of our series:

$$p_k := \begin{cases} a_k & \text{if } a_k \geq 0 \\ 0 & \text{if } a_k \leq 0 \end{cases} \quad \text{and} \quad n_k := \begin{cases} 0 & \text{if } a_k \geq 0 \\ -a_k & \text{if } a_k \leq 0. \end{cases}$$

Since absolute convergence implies convergence, the positive and negative series both must converge, and indeed

$$\sum_{k=0}^{\infty} p_k = \frac{1}{2}\left(\sum_{k=0}^{\infty} |a_k| + \sum_{k=0}^{\infty} a_k \right) \quad \text{and} \quad \sum_{k=0}^{\infty} n_k = \frac{1}{2}\left(\sum_{k=0}^{\infty} |a_k| - \sum_{k=0}^{\infty} a_k \right).$$

See Exercise 3.11 for details on how the above works. Because of the above, it is sufficient to show that a series where all of the terms are positive can be permuted without changing its value.

Let $\bar{\mu}$ be the permutation of the terms of convergent series $\sum_{k=0}^{\infty} p_k$ so that

$$p_{\bar{\mu}(1)} \geq p_{\bar{\mu}(2)} \geq p_{\bar{\mu}(3)} \geq \cdots \geq p_{\bar{\mu}(k)} \geq p_{\bar{\mu}(k+1)} \cdots.$$

This permutation must be possible since by the n-th term test, $\lim_{k\to\infty} p_k = 0$, meaning that only a finite number of terms can be greater than, say, 1/2, another finite number of terms can be between 1/2 and 1/4, and so forth. To see that this permuted series also converges, notice that its partial sums are bounded by certain of the partial sums of the original series:

$$\bar{S}_n := \sum_{k=0}^{n} p_{\bar{\mu}(k)} \leq S_{M(n)},$$

where $M(n) := \max\{\bar{\mu}(1), \bar{\mu}(2), \ldots \bar{\mu}(n)\}$. So $S_{M(n)}$ includes all of the n largest terms and likely some number of other terms too. Now as $n \to \infty$, also $S_{M(n)}$ increases to the value of the original series, $\sum_{k=0}^{\infty} p_k$, so $\bar{S}_n$ is an increasing sequence, bounded above by this value. Hence $\bar{S}_n$ converges, and so does the permuted series. Indeed, this argument shows that

$$\sum_{k=0}^{\infty} p_{\bar{\mu}(k)} \leq \sum_{k=0}^{\infty} p_k,$$

and since $S_n \leq \bar{S}_n$ (the sum of the first n terms cannot exceed the sum of the n largest terms),

$$\sum_{k=0}^{\infty} p_k \leq \sum_{k=0}^{\infty} p_{\bar{\mu}(k)}.$$

Thus,

$$\sum_{k=0}^{\infty} p_k = \sum_{k=0}^{\infty} p_{\tilde{\mu}(k)}.$$

Now observe that the previous argument would work if we started with any permutation of the original series. Thus,

$$\sum_{k=0}^{\infty} p_{\mu(k)} = \sum_{k=0}^{\infty} p_{\tilde{\mu}(k)}$$

for any permutation μ.

The proof of the first portion of this theorem (which is actually the portion proven by Riemann) is discussed in Exercise B.3.12. □

The Riemann–Dirichlet series theorem more or less completely answers the question of when the terms in a series can be reordered (permuted, commuted). The case of regrouping (reassociating) is more complicated. Regrouping terms (or splitting apart terms as in the next example) always changes the terms and therefore changes the partial sums; the question then is when it changes the limit value for the sequence of partial sums.

Example 3.4.1. Consider the following series:

$$3 - \frac{5}{2} + \frac{5}{2} - \frac{9}{4} + \frac{9}{4} - \frac{17}{8} + \cdots + \frac{2^{k+1}+1}{2^k} - \frac{2^{k+2}+1}{2^{k+1}} + \cdots.$$

Does this series converge, and if so, to which value? Does regrouping terms (without reordering) change the outcome?

Answer. As written, this series diverges by the n-th term test since the absolute value of each term is bigger than 2. But notice what happens when an odd–even grouping of terms occurs:

$$\left(3 - \frac{5}{2}\right) + \left(\frac{5}{2} - \frac{9}{4}\right) + \left(\frac{9}{4} - \frac{17}{8}\right) + \cdots = \frac{1}{2} + \frac{1}{4} + \frac{1}{8} + \cdots = 1.$$

With this regrouping, the series becomes an absolutely convergent geometric series. Of course, if regrouping (or degrouping) is allowed, then one could move backward from this absolutely convergent geometric series to the original divergent series. On the other hand, an even–odd grouping of terms yields

$$3 - \left(\frac{5}{2} - \frac{5}{2}\right) - \left(\frac{9}{4} - \frac{9}{4}\right) - \left(\frac{17}{8} - \frac{17}{8}\right) - \cdots = 3 - 0 - 0 - \cdots = 3.$$

All of this indicates that regrouping of terms in a series is complicated and dangerous.

Exercises 3, Group A

3.1. Does each of the following geometric series (if each really is geometric) converge or diverge? Justify your answer, and if it converges, state the value. In some cases, the series may be combinations of geometric series.

(a) $\displaystyle\sum_{n=0}^{\infty} \frac{1}{5^n}$

(e) $\displaystyle\sum_{k=2}^{\infty} (-2/5)^k$

(b) $\displaystyle\sum_{k=1}^{\infty} \frac{2}{5^k}$

(f) $\displaystyle\sum_{m=1}^{\infty} \frac{1}{1-(1/3)^m}$

(c) $\displaystyle\sum_{k=0}^{\infty} (-e/\pi)^k$

(g) $\displaystyle\sum_{k=0}^{\infty} \frac{3^k - 1}{5^k}$

(d) $\displaystyle\sum_{m=0}^{\infty} \frac{(-1)^m 7^m}{6^{m+3}}$

(h) $\displaystyle\sum_{n=1}^{\infty} \frac{(-2)^{n+1} - 3^n}{5^{n-1}}$

Answers. (a) Converges to 5/4. (b) Converges to 1/2. (c) Converges to $\frac{\pi}{\pi+e}$. (f) Actually not a geometric series; diverges; m-th term goes to one, not zero. (h) Converges to $-65/14$ (I think).

3.2. Does each of the following alternating series (if each really is alternating) converge or diverge? Justify your answer.

(a) $\displaystyle\sum_{k=0}^{\infty} \frac{(-1)^k}{k+1}$

(e) $\displaystyle\sum_{k=2}^{\infty} (-2/5)^k$

(b) $\displaystyle\sum_{k=0}^{\infty} \frac{(-1)^k k}{(k+2)^2}$

(f) $\displaystyle\sum_{m=0}^{\infty} \frac{(-1)^m}{\sqrt{m+1}}$

(c) $\displaystyle\sum_{k=1}^{\infty} \frac{(-1)^k k^3}{(3k+1)^3}$

(g) $\displaystyle\sum_{n=1}^{\infty} \sin\left(\frac{(-1)^n}{n}\right)$

(d) $\displaystyle\sum_{m=0}^{\infty} \frac{(-1)^m 7^m}{6^{m+1}}$

(h) $\displaystyle\sum_{n=1}^{\infty} \left(1 + \frac{(-1)^n}{n}\right)^n$

Answers. (a) Converges; the k-th term decreases to zero. (c) Diverges; the k-th term approaches 1/27. (f) Converges; the m-th term decreases to zero. (g) Converges; $\sin((-1)^n/n) = (-1)^n \sin(1/n)$, which alternates and decreases to zero.

3.3. Please determine the convergence or divergence of each of the following series using the comparison test.

(a) $\displaystyle\sum_{m=0}^{\infty} \frac{1}{m^2 + 3}$

(c) $\displaystyle\sum_{n=1}^{\infty} \frac{1}{n^2 + \sqrt{n}}$

(b) $\displaystyle\sum_{j=7}^{\infty} \frac{2}{j-5}$

(d) $\displaystyle\sum_{n=1}^{\infty} \frac{1}{n + \sqrt{n}}$

(e) $\displaystyle\sum_{\ell=1}^{\infty} \frac{2^{\ell}}{1+3^{\ell}}$ 　　　　　 (g) $\displaystyle\sum_{k=1}^{\infty} \frac{\ln k}{k^2}$

(f) $\displaystyle\sum_{k=1}^{\infty} \frac{\ln k}{k^3}$ 　　　　　 (h) $\displaystyle\sum_{i=1}^{\infty} \frac{\cos i}{1+i^2}$

Answers. (b) Diverges by comparison with the harmonic series. (c) Converges by comparison with p-series with $p = 2$. (f) Converges by comparison with p-series with $p = 2$.

3.4. Use the integral test to determine the convergence or divergence of each of the following series.

(a) $\displaystyle\sum_{j=1}^{\infty} \frac{1}{1+j}$ 　　　　　 (e) $\displaystyle\sum_{k=1}^{\infty} \frac{3}{1+4k^2}$

(b) $\displaystyle\sum_{k=1}^{\infty} \frac{k}{1+k^2}$ 　　　　　 (f) $\displaystyle\sum_{m=1}^{\infty} \frac{m}{1+m^4}$

(c) $\displaystyle\sum_{m=2}^{\infty} \frac{1}{m\ln m}$ 　　　　　 (g) $\displaystyle\sum_{n=1}^{\infty} ne^{-n}$

(d) $\displaystyle\sum_{n=1}^{\infty} \frac{\ln n}{n}$ 　　　　　 (h) $\displaystyle\sum_{i=1}^{\infty} \frac{\tan^{-1} i}{1+i^2}$

Answers. (a) Diverges; let $x = 1 + j$ and integrate from 2 to ∞. (c) Diverges; let $x = \ln m$ and integrate from $\ln 2$ to ∞. (e) Converges; let $x = 2k$ and integrate from 2 to ∞. (g) Converges; integrate by parts.

3.5. Use the ratio test or the root test to determine the convergence o

(a) $\displaystyle\sum_{k=1}^{\infty} \frac{3^k}{k!}$ 　　　　　 (d) $\displaystyle\sum_{n=1}^{\infty} (\ln 2)^n$

(b) $\displaystyle\sum_{j=0}^{\infty} \frac{j+2}{j!}$ 　　　　　 (e) $\displaystyle\sum_{n=1}^{\infty} n!e^{-n}$

(c) $\displaystyle\sum_{j=0}^{\infty} \left(\frac{5j^2+2}{2j^2+1}\right)^j$ 　　　　　 (f) $\displaystyle\sum_{i=1}^{\infty} \left(\frac{1}{5}+\frac{5}{i}\right)^i$

Answers. (a) Converges (ratio test). (c) Diverges (root test). (d) Converges (ratio test or root test), and indeed converges to $\ln 2/(1-\ln 2)$ (geometric series) since $\ln 2 < 1$ $(2 < e)$.

3.6. Using any appropriate test, please determine the convergence or divergence of each of the following series. Where appropriate, determine if convergence is absolute or conditional. When possible, please give the value that the series converges to.

(a) $\displaystyle\sum_{n=0}^{\infty} \frac{(-1)^n}{\sqrt{n}}$ 　　　　　 (b) $\displaystyle\sum_{n=0}^{\infty} \frac{(-6)^n}{n!}$

(c) $\displaystyle\sum_{k=1}^{\infty} \frac{2}{7^k}$

(l) $\displaystyle\sum_{\ell=2}^{\infty} \left(\frac{7\ell+3}{\ell^2-1}\right)^{\ell}$

(d) $\displaystyle\sum_{n=0}^{\infty} \frac{4n}{n^4+1}$

(m) $\displaystyle\sum_{n=0}^{\infty} \frac{1+3^n+7^n}{5^n}$

(e) $\displaystyle\sum_{j=0}^{\infty} \frac{2^j+5^j}{8^{(j+2)}}$

(n) $\displaystyle\sum_{k=0}^{\infty} \frac{(-1)^k}{2^k}$

(f) $\displaystyle\sum_{k=1}^{\infty} ke^{-k^2}$

(o) $\displaystyle\sum_{m=1}^{\infty} \sqrt[m]{2}$

(g) $\displaystyle\sum_{m=1}^{\infty} \left(1+\frac{4}{m}\right)^m$

(p) $\displaystyle\sum_{k=1}^{\infty} \sin^k 1$

(h) $\displaystyle\sum_{n=1}^{\infty} \frac{n^2-2n+4}{n^3+5n-1}$

(q) $\displaystyle\sum_{i=1}^{\infty} \frac{i}{e^{i^2}}$

(i) $\displaystyle\sum_{i=2}^{\infty} \frac{\ln i}{i^2}$

(r) $\displaystyle\sum_{n=2}^{\infty} \frac{(-1)^{n+1}n}{\ln n}$

(j) $\displaystyle\sum_{j=0}^{\infty} \frac{1-\cos(j\pi/3)}{2^j}$

(s) $\displaystyle\sum_{\ell=1}^{\infty} \sin\left(\frac{1}{\ell}\right)$

(k) $\displaystyle\sum_{k=1}^{\infty} \frac{\sin(k\pi/2)}{\ln(2+k)}$

(t) $\displaystyle\sum_{m=2}^{\infty} (-7)^{5-m}$

Answers. (b) Converges (alternating series test), and indeed converges absolutely (ratio test). (c) Geometric series; converges to 1/3. (g) Diverges; the *n*-th (or *m*-th) term does not to go zero. (k) Converges conditionally. (m) Diverges by comparison with a divergent geometric series. (o) Diverges by the *n*-th term test. (t) Geometric series; converges to −2401/8. (I hope.)

Exercises 3, Group B

General advice: Consider each series as the limit of a sequence of partial sums.

3.1. Please prove that the series

$$\sum_{k=1}^{\infty} a_k = 1-1+\frac{1}{3}-\frac{1}{3}+\frac{1}{5}-\frac{1}{5}+\frac{1}{7}-\frac{1}{7}+\cdots$$

converges to zero.

Hint: Write S_n piecewise depending on whether n is even or odd.

3.2. Consider the function

$$f(x) = \sum_{k=0}^{\infty} x^k = 1 + x + x^2 + \cdots.$$

What is the domain (as a subset of $\mathbb{R}$) for this function? That is, for which $x \in \mathbb{R}$ does this series converge? Prove/explain your answer.

Hint: For $|x| \geq 1$, consider the n-th term test.

3.3. Suppose that $\{a_k\}$ is a sequence of real numbers and that $b_k = a_k - a_{k-1} \,\forall\, k \in \mathbb{Z}^+$.
(a) Prove that the series $\sum_{k=1}^{\infty} b_k$ converges iff the sequence $\{a_k\}$ converges.
(b) Let the sequence $\{a_k\}$ converge to L; evaluate the series $\sum_{k=1}^{\infty} b_k$.
(c) Suppose that $c_k = (a_k + a_{k-1})(a_k - a_{k-1}) \,\forall\, k \in \mathbb{Z}^+$. If again $\{a_k\}$ converge to L, evaluate the series $\sum_{k=1}^{\infty} c_k$.

3.4. If the series $\sum_{n=0}^{\infty} a_n$ converges, please show that the series $\sum_{n=0}^{\infty} 3a_n$ also converges and that indeed

$$\sum_{n=0}^{\infty} 3a_n = 3 \sum_{n=0}^{\infty} a_n.$$

Hint: The equality in this exercise seems obviously true, but how does one know for sure—how can it be proven? The result is trivial for finite sums, but can this be used to prove this result for series? Write the series as the limit of the sequence of partial sums and use the limit operations theorem to move the factor 3 through the limit.

3.5. Suppose that $\sum_{k=1}^{\infty} a_k$ converges.
(a) Suppose $a_k > 0$. Please prove that the series $\sum_{k=1}^{\infty} a_k^2$ also converges.
(b) Please give an example to show that some assumption like $a_k > 0$ is needed. Consider an appropriate alternating series.

3.6. Consider the series

$$\sum_{n=1}^{\infty} \frac{1}{2n-1} - \frac{1}{(2n)^2} = 1 - \frac{1}{2^2} + \frac{1}{3} - \frac{1}{4^2} + \frac{1}{5} - \frac{1}{6^2} + \cdots.$$

This is an alternating series where the n-th term goes to zero as $n \to \infty$. Does this series converge or diverge? Please explain why or why not.

Hint: Consider the partial sums for this series.

3.7. Please give examples of series where

$$\lim_{n \to \infty} \left| \frac{a_{n+1}}{a_n} \right| = 1$$

and

(a) the series converges absolutely.
(b) the series converges conditionally.
(c) the series diverges.
      Hint: Think about p-series.

3.8. Consider the proof that absolute convergence implies convergence (Theorem 3.3.3) and explain why it *cannot* be used to show the converse—that convergence of a series with both positive and negative terms implies absolute convergence.

3.9. Prove that the ratio test implies divergence when $r > 1$.
      Hint: Follow the plan of the convergence proof in the text when $r < 1$. In particular, define ρ as the midpoint as in the convergence proof, but now compare the given series to the appropriate divergent geometric series.

3.10. Prove that the root test implies divergence when $r > 1$.
      Hint: Follow the plan of the convergence proof in the text when $r < 1$. In particular, define ρ as the midpoint as in the convergence proof, but now compare the given series to the appropriate divergent geometric series.

3.11. In the proof of the Riemann–Dirichlet series theorem (Theorem 3.4.1), use partial sums and the monotone convergence theorem to show that representations for

$$\sum_{k=0}^{\infty} p_k \quad \text{and} \quad \sum_{k=0}^{\infty} n_k$$

are correct.

3.12. Please prove the first portion of the Riemann–Dirichlet series theorem, Theorem 3.4.1.
      Hint: For any $x \in \mathbb{R}^+$, show that a given conditionally convergent series can be reordered into three series, the first converging to x and the latter two canceling each other, so that the entire permuted series converges to x.

4 Sequences and series of functions

Up to this point, we have considered almost exclusively either sequences of real numbers $\{a_n\}$ or series of real numbers $\sum_{n=1}^{\infty} a_n$. These are important in and of themselves, but they are not sufficient for the study of calculus. Now we turn our attention to sequences and series of functions, and this will allow us to introduce and study calculus.

4.1 Pointwise versus uniform convergence

As it turns out, there are many types of convergence for sequences of functions; some are far beyond the scope of this text. Two of the most important ones, however, are also important for our present work. The first is pointwise convergence; this is most similar to the convergence of sequences of real numbers. The second is uniform convergence; this is a much more powerful type of convergence. These are considered in turn, first looking at pointwise convergence, and then studying uniform convergence.

Definition. Let $\mathcal{D} \subset \mathbb{R}$ be a domain in the real number line ($\mathcal{D}$ is often but not always a closed interval $[a, b]$ between two real numbers a and b). A sequence of functions $f_n : \mathcal{D} \to \mathbb{R}$ **converges pointwise** to a function $f : \mathcal{D} \to \mathbb{R}$ on the domain $\mathcal{D}$ iff given any $\epsilon > 0$ and any $x_0 \in \mathcal{D}$, $\exists N \in \mathbb{Z}^+$ such that

$$|f_n(x_0) - f(x_0)| < \epsilon$$

whenever $n > N$.

Remark. Notice that this definition assumes that both x_0 and ϵ are fixed at the start of finding the limit, and with these fixed, a value for N is obtained. If either x_0 or ϵ is changed, a new (perhaps larger) value of N must be found. Because x_0 is fixed, despite the fact that we started with a sequence of functions f_n, we are now looking at nothing more than a sequence of real numbers: $a_n = f_n(x_0)$. As such, all of the results from Chapter 1, *Sequences in* $\mathbb{R}$, can now be put to work for us here. So rather than requiring us to learn how to handle a new kind of convergence, pointwise convergence lets us use what we already know.

Example 4.1.1. Consider the sequence of functions $f_n(x) = x^n$ defined on the domain $\mathcal{D} = [0, 1]$. Does this sequence converge pointwise to a limit, and if so, what is that limit?

Answer. Notice that there are two points in the interval $[0, 1]$ where convergence is trivial: the endpoints. At $x = 0$, $f_n(0) = 0 \; \forall n \in \mathbb{Z}^+$, and at $x = 1$, $f_n(1) = 1 \; \forall n \in \mathbb{Z}^+$. So at $x = 0$, the sequence $\{f_n(0)\} = \{0, 0, 0, \ldots\}$, which of course must converge to a limit value $L = f(0) = 0$. Similarly, the sequence $\{f_n(1)\}$ converges to a limit value $L = f(1) = 1$. But now, what happens in the middle? Suppose that x_0 is some fixed value in $(0, 1)$. Then

https://doi.org/10.1515/9783110768398-004

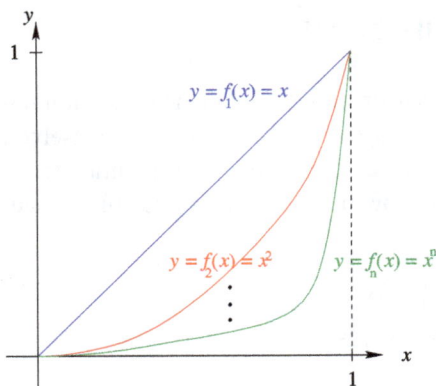

Figure 4.1: The sequence of functions $f_n(x) = x^n$ defined on the domain $\mathcal{D} = [0,1]$. Notice that this sequence converges pointwise to $f(x) = 0$ for $x \in [0,1)$, but $f(1) = 1$. So in this case, the limit function f is discontinuous.

$f_n(x_0) = x_0^n$ and

$$\lim_{n\to\infty} f_n(x_0) = \lim_{n\to\infty} x_0^n = 0$$

since for $0 < x_0 < 1$, raising x_0 to higher and higher powers makes it smaller and smaller (though still positive). This limiting process is shown in Figure 4.1. So this sequence of functions does converge pointwise to a limit on $[0,1]$, and the limit function f is

$$f(x) = \begin{cases} 0 & x \in [0,1) \\ 1 & x = 1. \end{cases}$$

Remark. Notice that in the example above, each of the functions f_n is continuous on $[0,1]$, while the limit function f is *not* continuous. This is in fact a hallmark of pointwise convergence: A sequence of continuous functions may converge to a discontinuous limit. This is also a significant weakness; one would prefer that sequences of continuous functions converge to continuous limit functions. Our second form of convergence (uniform convergence) has the huge advantage that it preserves continuity: If a sequence of continuous functions converges uniformly to a limit, that limit *must* be continuous.

Definition. Let $\mathcal{D} \subset \mathbb{R}$ be a domain in the real number line (again, $\mathcal{D}$ is often but not always an interval $[a,b]$ between two real numbers a and b). A sequence of functions $f_n : \mathcal{D} \to \mathbb{R}$ **converges uniformly** to a function $f : \mathcal{D} \to \mathbb{R}$ on the domain $\mathcal{D}$ iff given any $\epsilon > 0$, $\exists N \in \mathbb{Z}^+$ such that

$$|f_n(x) - f(x)| < \epsilon \quad \forall x \in \mathcal{D}$$

whenever $n > N$.

Remark. Notice that the definition of uniform convergence differs from the definition of pointwise convergence in that ϵ is still fixed at the start of finding the limit, but the same value of N must now work for any and all $x \in \mathcal{D}$. If ϵ is decreased, probably a new (larger) value of N must be found, but it must continue to work for all $x \in \mathcal{D}$.

Example 4.1.2. Consider the sequence of functions $f_n(x) = x/n$ defined on the domain $\mathcal{D} = [0, 1]$. Does this sequence converge (1) pointwise and (2) uniformly to a limit, and if so, what is that limit?

Remark. Notice that by definition, if a sequence of functions converges uniformly, then it must also converge pointwise to the same limit function. So a general strategy for considering uniform convergence is to first find the pointwise limit and then ask if it is also the uniform limit.

Answer. Using the strategy mentioned above, one should first find the pointwise limit. Pick $x_0 \in \mathcal{D} = [0, 1]$ and consider

$$\lim_{n \to \infty} f_n(x_0) = \lim_{n \to \infty} x_0/n.$$

Notice that since $0 < x_0 < 1$, in the above limit, $0 < x_0/n < 1/n$. Thus, by the sandwich theorem,

$$\lim_{n \to \infty} f_n(x_0) = \lim_{n \to \infty} x_0/n = 0.$$

Since f_n converges pointwise to 0, the question to ask now is whether, given $\epsilon > 0$, there is an $N \in \mathbb{Z}^+$ such that

$$\left| f_n(x) - 0 \right| = |x/n - 0| = x/n < \epsilon \quad \forall x \in [0, 1]$$

whenever $n > N$. But again $x/n < 1/n \, \forall x \in [0, 1]$, so we simply need $1/n < \epsilon \Leftrightarrow n > 1/\epsilon$. Hence the definition of uniform convergence is satisfied for this example if $N := \lceil 1/\epsilon \rceil$, and this means that for our current example, f_n converges uniformly to 0. This uniform convergence is shown in Figure 4.2.

Example 4.1.2 above may make it seem as though there is little difference between pointwise convergence and uniform convergence. While this may be true for the previous example, it is not true in general, and the next example, along with Example 4.1.1, shows how much difference there is.

Example 4.1.3. Consider the sequence of functions $f_n(x) = \frac{nx}{1+n^2x^2}$ defined on the domain $\mathcal{D} = [0, 1]$. Show that this sequence converges pointwise, but not uniformly. What is the pointwise limit?

Answer. First, considering pointwise convergence, fix $x = x_0 \in [0, 1]$ and compute

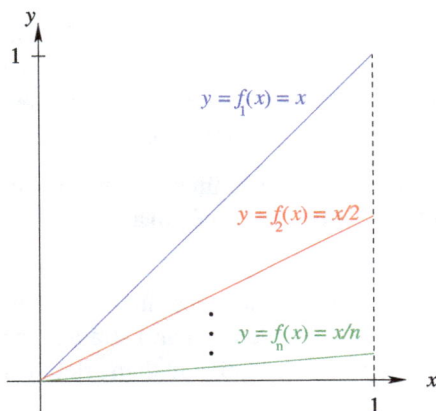

Figure 4.2: The sequence of functions $f_n(x) = x/n$ defined on the domain $\mathcal{D} = [0,1]$. Notice that this sequence converges uniformly to $f(x) \equiv 0$.

$$\lim_{n \to \infty} f_n(x_0) = \lim_{n \to \infty} \frac{nx_0}{1 + n^2 x_0^2} = \lim_{n \to \infty} \frac{x_0}{1/n + nx_0^2} = 0$$

($x_0 = 0$ and $x_0 > 0$ are separate cases, but the result is the same either way).

So if this sequence $\{f_n\}$ converges uniformly to anything, it must converge uniformly to 0. But to see that this is not the case, one can find the location of the maximum value of f_n on $[0,1]$ for each n. Notice that $f_n(0) = 0 \ \forall n \in \mathbb{Z}^+$. Since $f_n(x) > 0 \ \forall x \in (0,1]$, if there is a single value of x_M where $f_n'(x_M) = 0$, then f_n must achieve its maximum value at that x_M. Computing the derivative of f_n with respect to x using the quotient rule and setting $f_n'(x) = 0$, one finds

$$f_n'(x) = \frac{n - n^3 x^2}{(1 + n^2 x^2)^2} = 0.$$

For any rational expression, the only way that it can be zero is if its numerator is zero. Moreover, since $n \in \mathbb{Z}^+$, an n can be factored out of the numerator, implying that $f_n'(x) = 0$ for $0 < x < 1$ only when $x = 1/n$. Now notice that $f_n(0) = 0 \ \forall n \in \mathbb{Z}^+$, $f_n(x) > 0$ on $(0,1] \ \forall n \in \mathbb{Z}^+$, and

$$\lim_{n \to \infty} f_n(1) = \lim_{n \to \infty} \frac{n}{1 + n^2} = 0.$$

All of this means that each f_n has its maximum value at $x = 1/n$. By direct evaluation, one sees that $f_n(1/n) = 1/2 \ \forall n \in \mathbb{Z}^+$. So each of the functions in this sequence has the same maximum value, 1/2, but this maximum value is achieved at a different point in the domain for each n. Thus this sequence of functions converges pointwise to zero, but cannot converge uniformly to anything because each element of the sequence pops up to 1/2 before falling back to zero. This sequence is shown in Figure 4.3. Notice that since

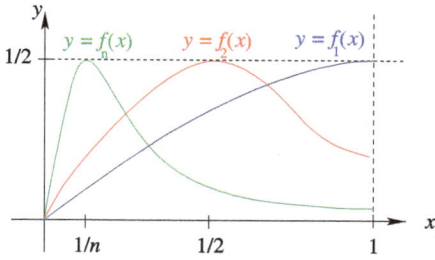

Figure 4.3: The sequence of functions $f_n(x) = \frac{nx}{1+n^2x^2}$ defined on the domain $\mathcal{D} = [0,1]$. This sequence converges pointwise, but not uniformly, to $f(x) \equiv 0$. In particular, $\forall n \in \mathbb{Z}^+, f_n(1/n) = 1/2$.

the location of the maximum is at $x = 1/n$, this location gets closer and closer to zero as n increases.

The previous example is probably one of the best to show the difference between converging to zero pointwise and converging to zero uniformly. The next is similar.

Example 4.1.4. For $n > 1$, consider the sequence of functions

$$f_n(x) = \begin{cases} n^2 x & 0 \le x \le 1/n \\ 2n - n^2 x & 1/n \le x \le 2/n \\ 0 & 2/n \le x \le 1 \end{cases}$$

defined on the domain $\mathcal{D} = [0,1]$. Show that this sequence converges pointwise, but not uniformly. What is the pointwise limit? What is that integral of f_n?

Answer. This sequence of tent functions is shown in Figure 4.4. Notice that for each $x_0 \in (0,1)$, $f_n(x_0) = 0$ when $n > 2/x_0$ since for n that large, x_0 is in the portion of the domain to the right of the entire tent. Hence, once again $f_n \to f \equiv 0$ pointwise. But notice that $f_n(1/n) = n$, so the maximum value of f_n in fact blows up (goes to $+\infty$) as n increases. In fact,

$$\int_0^1 f_n(x)\,dx = 1 \quad \forall n \in \mathbb{Z}^+ - \{1\},$$

while the integral of the limit is very different:

$$\int_0^1 f(x)\,dx = \int_0^1 0\,dx = 0.$$

This gets at another flaw in pointwise convergence: The value of the integral of f_n tells us nothing about the value of the integral of the sequence limit, f.

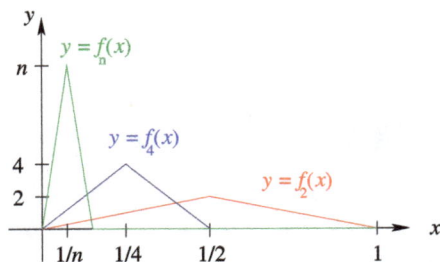

Figure 4.4: The sequence of tent functions given in Example 4.1.4. This sequence converges pointwise, but not uniformly, to $f(x) \equiv 0$. For each n, there is unit area under the tent.

The examples above now guide our discussions of why uniform convergence is superior to pointwise convergence. The next section presents two theorems on this topic.

4.2 Implications of uniform convergence

There are a number of important implications of uniform convergence—two were already mentioned above: A uniformly convergent sequence of continuous functions always converges to a continuous function, and the limit of the integrals of a sequence of functions converges to the integral of the limit of the sequence. These results are considered next.

Theorem 4.2.1. *Suppose that $f_n \to f$ uniformly on some interval $[a,b]$ and that each f_n is continuous. Then f must also be continuous.*

Remark. As mentioned in Example 4.1.1, this theorem has an important consequence when a sequence of continuous functions converges to a limit that is not continuous. This theorem says that such convergence cannot be uniform.

Proof. From the definition of continuous, given $\epsilon > 0$, we must find a $\delta > 0$ such that if $x, y \in [a,b]$ and $|x-y| < \delta$, then $|f(x)-f(y)| < \epsilon$. But notice that by the triangle inequality,

$$|f(x) - f(y)| \le |f(x) - f_n(x)| + |f_n(x) - f_n(y)| + |f_n(y) - f(y)|.$$

Since $f_n \to f$ uniformly, there exists $N \in \mathbb{Z}^+$ such that both $|f(x) - f_n(x)| < \epsilon/3$ and $|f(y) - f_n(y)| = |f_n(y) - f(y)| < \epsilon/3$ for all $n > N$. Pick and fix one such $n > N$. Now we consider the middle term in the sum. Since f_n is continuous, there exists $\delta > 0$ such that $|f_n(x) - f_n(y)| < \epsilon/3$ whenever $|x - y| < \delta$. Thus,

$$|f(x) - f(y)| \le |f(x) - f_n(x)| + |f_n(x) - f_n(y)| + |f_n(y) - f(y)|$$
$$< \quad \epsilon/3 \quad + \quad \epsilon/3 \quad + \quad \epsilon/3 \quad = \quad \epsilon$$

provided that $n > N$ and $|x - y| < \delta$. This of course is exactly what was to be proven. ☐

While the previous result is helpful in understanding the difference between point-wise and uniform convergence, the next result is essential in understanding how to carry out calculus on sequences and series of functions, particular for the power series discussed below. The result says, roughly speaking, that when a sequence of functions converges uniformly, the limit of the integral of the sequence is the same as the integral of the limit of the sequence. Before discussing this result, the following definition is helpful.

Definition. If f is a real-valued function defined on the interval $[a, b]$ (so $f : [a, b] \to \mathbb{R}$), then f is **Riemann[1] integrable** on $[a, b]$ iff

$$\int_a^b f(x)\, dx$$

exists as a finite real number (*not* $\pm\infty$). In symbols, a function f being Riemann integrable on $[a, b]$ is denote by $f \in R[a, b]$, the letter "R" of course standing for "Riemann".

Theorem 4.2.2 (Uniform convergence theorem). *Suppose* $f_n \to f$ *uniformly with* $f_n \in R[a, b]$. *Then* $f \in R[a, b]$ *and*

$$\lim_{n\to\infty} \int_a^b f_n(x)\, dx = \int_a^b \left(\lim_{n\to\infty} f_n(x) \right) dx = \int_a^b f(x)\, dx.$$

Proof. First we will show that $\lim_{n\to\infty} \int_a^b f_n(x)\, dx$ exists as a real number by showing that the sequence $\{\int_a^b f_n(x)dx\}$ is a Cauchy sequence of real numbers (recall that by Theorem 2.2.1, Cauchy sequences of real numbers always converge to real numbers). Given $\epsilon > 0$, since $f_n \to f$ uniformly, then $\exists N \in \mathbb{Z}^+$ such that $\forall m, n > N$, $|f_n(x) - f_m(x)| < \epsilon/(b - a)\ \forall x \in [a, b]$. So

$$\left| \int_a^b f_n(x)\, dx - \int_a^b f_m(x)\, dx \right| = \left| \int_a^b (f_n(x) - f_m(x))\, dx \right|$$

$$\leq \int_a^b |f_n(x) - f_m(x)|\, dx$$

$$< \int_a^b \frac{\epsilon}{b - a}\, dx = \epsilon.$$

---

1 Bernhard Riemann (1826–1866) was a German mathematician who worked in analysis, number theory and differential geometry and who is well known for developing the first rigorous treatment of integration.

Thus, $\{\int_a^b f_n(x)dx\}$ is a Cauchy sequence of real numbers, which implies that $\{\int_a^b f_n(x)dx\}$ converges to some real number. That is, $\exists L \in \mathbb{R}$ such that

$$\lim_{n\to\infty} \int_a^b f_n(x)\, dx = L.$$

So our sequence of integrals converges to some real number L. Now we must show that $f \in R[a,b]$ and that $L = \int_a^b f(x)dx$. To accomplish both these goals, consider the Riemann sums for the integral of f. We need to show that $|L - \sum_P f(\xi_i)\Delta x_i|$ is sufficiently small for any partition P that is sufficiently fine. The main tool in accomplishing this is again the triangle inequality:

$$\left| L - \sum_P f(\xi_i)\Delta x_i \right|$$

$$\leq \left| L - \int_a^b f_n(x)dx \right| + \left| \int_a^b f_n(x)dx - \sum_P f_n(\xi_i)\Delta x_i \right| + \left| \sum_P f_n(\xi_i)\Delta x_i - \sum_P f(\xi_i)\Delta x_i \right|.$$

Now we need to show that each of the three terms on the right side of the inequality is small. So given $\epsilon > 0$, since $\lim_{n\to\infty} \int_a^b f_n(x)dx = L$ from the first part above, $\exists N_1 \in \mathbb{Z}^+$ such that $\forall n > N_1$, $|L - \int_a^b f_n(x)dx| < \epsilon/3$. Then since $f_n \to f$ uniformly, $\exists N_2 \in \mathbb{Z}^+$ such that $\forall n > N_2$, $|f_n(x) - f(x)| < \epsilon/3(b-a)$ $\forall x \in [a,b]$. Thus, for the third part,

$$\left| \sum_P f_n(\xi_i)\Delta x_i - \sum_P f(\xi_i)\Delta x_i \right| = \left| \sum_P (f_n(\xi_i) - f(\xi_i))\Delta x_i \right|$$

$$\leq \sum_P |f_n(\xi_i) - f(\xi_i)|\Delta x_i$$

$$< \sum_P \left(\frac{\epsilon}{3(b-a)} \right)\Delta x_i$$

$$= \frac{\epsilon}{3(b-a)} \sum_P \Delta x_i$$

$$= \frac{\epsilon}{3(b-a)}(b-a) = \frac{\epsilon}{3}.$$

Finally, fix $n > \max\{N_1, N_2\}$ so that both of these first two $\epsilon/3$ bounds hold. Since $f_n \in R[a,b]$, $\exists \delta > 0$ such that $\forall$ partitions P with $\|P\| < \delta$, $|\int_a^b f_n(x)dx - \sum_P f_n(\xi_i)\Delta x_i| < \epsilon/3$. Therefore, each of the three terms can be made less than $\epsilon/3$ by moving far down the sequence so that $n > \max\{N_1, N_2\}$ and choosing the partition with $\|P\| < \delta$. The proof is complete. $\qquad\square$

So uniform convergence is sufficient to allow the exchange of limits for integrals of sequences (or sequences of integrals), but as was seen in Example 4.1.4, pointwise convergence is not sufficient. One might wonder whether uniform convergence is required for this sort of exchange of limits. The answer is that it is not. There are in fact a number of convergence theorems (dominated convergence theorem, monotone convergence theorem for integrals), but these are beyond the scope of this text. For our discussion here, the uniform convergence theorem is sufficient.

The uniform convergence theorem is most useful when one side of the equality (often the integral of the limit) is relatively easy to compute, but the other side is much harder. In the following example, both sides of the equality can be computed, so one can see that they are the same.

Example 4.2.1. Consider the sequence of functions $f_n(x) = \sin x + (1/n)\cos(x/n)$ defined on the interval $\mathcal{D} = [0, \pi]$. Find the limit of the integral of f_n:

$$\lim_{n \to \infty} \int_0^\pi f_n(x)\, dx.$$

Answer. To use the uniform convergence theorem, one needs $f_n \to f$ uniformly for some f. In this case, it would appear that $f(x) = \sin x$ should be the value of f. Indeed,

$$|f(x) - f_n(x)| = \left|\sin x - \sin x - \frac{1}{n}\cos(x/n)\right| = \frac{1}{n}|\cos(x/n)| \le \frac{1}{n}$$

$\forall x \in [0, \pi]$. Thus, the uniform convergence theorem applies and

$$\lim_{n \to \infty} \int_0^\pi f_n(x)\, dx = \int_0^\pi \left(\lim_{n \to \infty} f_n(x)\right) dx = \int_0^\pi f(x)\, dx = \int_0^\pi \sin x\, dx = 2.$$

But in this case, the integral of f_n can also be computed directly using the substitution $u = x/n$:

$$\int_0^\pi f_n(x)\, dx = \int_0^\pi \sin x + \frac{1}{n}\cos(x/n)\, dx = 2 + \sin(\pi/n).$$

Taking the limit as $n \to \infty$, one finds that the value found by computing the limit directly is the same as that given by the uniform convergence theorem, as of course it must.

In the previous example, the $(1/n)\cos(x/n)$ term can be thought of as a perturbation of the base function, which in this case is $f(x) = \sin x$. When dealing with perturbations, it is often known that the perturbation is small without knowing specifically what the perturbation is. In the following example, the integral of f_n cannot be computed explicitly, but the uniform convergence theorem still allows us to compute the limit of these integrals.

Example 4.2.2. Consider the sequence of functions $f_n(x) = \sqrt{|x|} + \epsilon_n(x)$ defined on the interval $\mathcal{D} = [-1, 1]$, where $\forall n \in \mathbb{Z}^+$, the function ϵ_n is continuous and bounded by $1/n$.

Find the limit of the integral of f_n,

$$\lim_{n \to \infty} \int_{-1}^{1} f_n(x)\, dx.$$

Answer. Notice that since e_n is continuous, it must be integrable on $[-1, 1]$; it is just that the value of such integrals cannot be explicitly determined. But of course in this case f_n converges uniformly to $f(x) = \sqrt{|x|}$, so by the uniform convergence theorem,

$$\lim_{n \to \infty} \int_{-1}^{1} f_n(x)\, dx = \int_{-1}^{1} \left(\lim_{n \to \infty} f_n(x) \right) dx = \int_{-1}^{1} \sqrt{|x|}\, dx = \frac{4}{3}.$$

As we shall see, the uniform convergence theorem is critical to doing calculus on an important series of functions, power series.

4.3 Power series

Perhaps surprisingly, in one of the most important series of functions, the functions are about as simple[2] as they can be: $f_n(x) = x^n$. Series where the terms all have the form a coefficient times a power of x are called power series (centered at zero).

Definition. Power series (centered at zero): A power series is any series of the form

$$\sum_{n=0}^{\infty} a_n x^n = a_0 + a_1 x + a_2 x^2 + a_3 x^3 + \cdots,$$

where the coefficients a_n are real numbers.

Perhaps not surprisingly, the most important question to ask about power series is: For which real values of x do these series converge? There is one trivial case that should be mentioned immediately: When $x = 0$, every term in the series is zero except for the zeroth term, so the entire series is simply a_0 in this case. The answer for more general values of x is still relatively simple provided that a certain limit involving the coefficients exists.

Theorem 4.3.1. *Suppose that the limit*

$$\lim_{n \to \infty} \frac{|a_{n+1}|}{|a_n|} = L,$$

---

2 Many authors prefer to consider functions of the form $f_n(x) = (x - x_0)^n$ at the start; we will broaden our definition to cover this case shortly.

where either $L \geq 0$ (a finite, nonnegative real number) or $L = +\infty$. Let $R := 1/L$ when L is finite and nonzero; take $R = 0$ when $L = +\infty$ and $R = +\infty$ when $L = 0$. Then the power series

$$\sum_{n=0}^{\infty} a_n x^n = a_0 + a_1 x + a_2 x^2 + a_3 x^3 + \cdots$$

converges absolutely for x in the interval $(-R, R)$ and diverges for $x < -R$ or $x > R$. Moreover, if $R_1 < R$, then the series converges uniformly for x in the interval $[-R_1, R_1]$. The parameter R is called the **radius of convergence** *and the interval $(-R, R)$ is called the* **interval of convergence**.

A schematic of the interval of convergence is shown in Figure 4.5. Before discussing the proof for this theorem, let us consider one example—one that we have seen before.

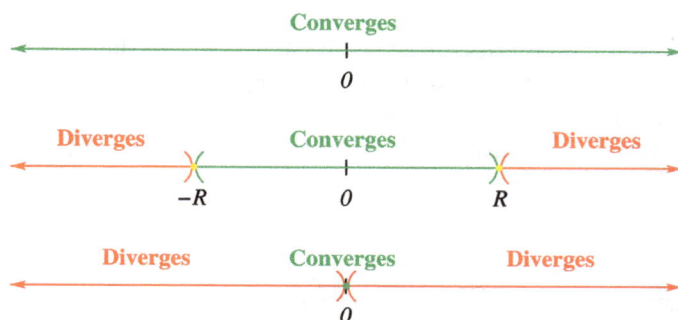

Figure 4.5: Schematic of the interval of convergence. In the middle case, $R > 0$ is finite, and the interval of convergence is shown in green. The semi-infinite intervals of divergence are shown in red. The endpoints at $-R$ and R are shown in yellow since neither convergence nor divergence is guaranteed at these points. In the top case, $R = +\infty$, so the interval of convergence is the entire real line. In the bottom case, $R = 0$, so convergence only occurs at a single point, $x = 0$.

Example 4.3.1. Consider the power series where $a_n = 1 \; \forall \, n \in \mathbb{Z}^+$:

$$\sum_{n=0}^{\infty} x^n = 1 + x + x^2 + x^3 + \cdots.$$

For which x does this power series converge, and for which does it diverge?

Answer. According to Theorem 4.3.1 above, since

$$\lim_{n \to \infty} \frac{|a_{n+1}|}{|a_n|} = 1,$$

this power series must be absolutely convergent for $-1 < x < 1$. But in fact, we already know that this is true, since this power series is a geometric series where the ratio between terms is x and the initial term is 1. Hence,

$$\sum_{n=0}^{\infty} x^n = \frac{1}{1-x}$$

for $-1 < x < 1$ as we saw before. By the n-th term test, this power series must diverge when $|x| \geq 1$ just as our current theorem guarantees, at least when $|x| > 1$.

Proof. The proof of the first part of the theorem (that the power series converges on the interval $(-R, R)$ and diverges outside the closed interval $[-R, R]$) is based on the ratio test and is fairly straightforward (read "easy"). Consider the limit of the ratio of two consecutive terms in this power series:

$$\lim_{n \to \infty} \frac{|a_{n+1} x^{n+1}|}{|a_n x^n|} = \lim_{n \to \infty} \frac{|a_{n+1}|}{|a_n|} |x| = L|x|.$$

According to the ratio test, our power series converges absolutely when the limit above is less than 1 and diverges when this limit is greater than 1. So our power series converges when $|x| < 1/L = R$ and diverges when $|x| > 1/L = R$.

The proof that convergence is uniform on any slightly smaller interval $[-R_1, R_1]$ is left as Exercise B.4.7. □

Remarks.

1. Notice that this theorem says nothing about what happens at the endpoints, and indeed, there is no general result for what happens at endpoints. If one wishes to know whether or not a power series converges at each of the endpoints of the interval of convergence, one must test each endpoint separately using one of the convergence tests discussed in the previous chapter. The good news is that in general, convergence at endpoints is not a key issue. Since we are interested in doing calculus on power series, we are much more interested in points inside the interval of convergence, not the endpoints where the series necessarily diverges on one side or the other. The endpoints are right on the ragged edge—never a good place for calculus.

2. One might think that our power series convergence theorem, Theorem 4.3.1, applies to every power series. This is not quite true. The theorem hinges on a certain limit converging. Should the limit of consecutive coefficients neither converge or diverge to infinity (for example, if the ratio alternates between two values), our power series convergence theorem does not apply, at least not directly; see Exercise B.4.8. Still, when this limit is finite or infinity, the theorem applies and yields the radius of convergence, and when this limit does not exist, there is still an interval of convergence as depicted in Figure 4.5, the radius R is just harder to determine; see Exercise B.4.9.

3. The general form for a power series is

$$\sum_{n=0}^{\infty} a_n (x - x_0)^n.$$

Converges

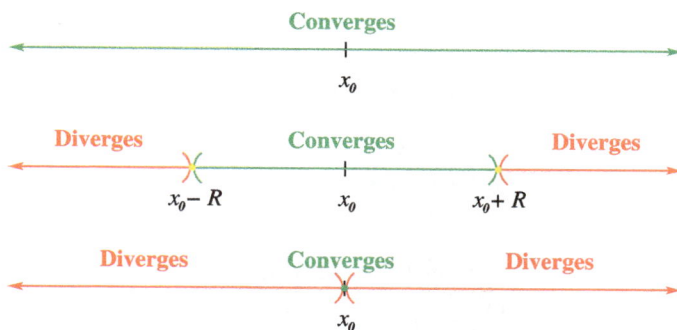

Diverges   Converges   Diverges

Diverges   Converges   Diverges

Figure 4.6: Schematic of the interval of convergence for $x_0 \neq 0$. In the middle case, $R > 0$ is again finite, and the interval of convergence is again shown in green; the semi-infinite intervals of divergence are shown in red; the endpoints in yellow are shifted to $x_0 - R$ and $x_0 + R$. The $R = 0$ and $R = +\infty$ cases are as before, except the center is now x_0.

Until now, we have taken $x_0 = 0$ to simplify our presentation. Notice that changing the value of x_0 does not change the radius of convergence; it simply shifts the interval of convergence. A schematic of this more general interval of convergence is shown in Figure 4.6. The parameter x_0 is called the **center** of the interval of convergence. For many of the power series considered next, $x_0 \neq 0$.

4.4 Calculus on power series

Suppose a function $f : (x_0 - R, x_0 + R) \to \mathbb{R}$ is defined as a power series

$$f(x) := \sum_{n=0}^{\infty} a_n (x - x_0)^n$$

for x in the interval of convergence. Is it possible to differentiate and integrate f? The answer is "Yes!" The next theorem tells us how.

Theorem 4.4.1. *Consider a function $f : (x_0 - R, x_0 + R) \to \mathbb{R}$ defined as a power series*

$$f(x) := \sum_{k=0}^{\infty} a_k (x - x_0)^k$$

for $x \in (x_0 - R, x_0 + R)$, the interval of convergence. Then f is both differentiable and Riemann integrable on the interval of convergence, and differentiation and integration can be performed term-by-term:

$$f'(x) = \sum_{k=1}^{\infty} k a_k (x - x_0)^{k-1} = \sum_{n=0}^{\infty} (n + 1) a_{n+1} (x - x_0)^n$$

and

$$\int_{x_0}^{x} f(\xi)\,d\xi = \sum_{k=0}^{\infty} \frac{a_k}{k+1}(x-x_0)^{k+1} = \sum_{n=1}^{\infty} \frac{a_{n-1}}{n}(x-x_0)^n.$$

The series for the derivative and those for the integral are the same except for an index shift: $n = k + 1$.

Proof. For the integration result, notice that since f is defined as a uniformly convergent power series on $[x_0 - R_1, x_0 + R_1]$ for any positive $R_1 < R$, it must be continuous as a function of x on $[x_0 - R_1, x_0 + R_1]$ by Theorem 4.2.1. Since R_1 can be chosen arbitrarily close to R, this series must be continuous on $(x_0 - R, x_0 + R)$ and therefore also Riemann integrable on any closed subinterval in $(x_0 - R, x_0 + R)$. Hence, for any $x \in (x_0 - R, x_0 + R)$

$$\int_{x_0}^{x} f(\xi)\,d\xi = \int_{x_0}^{x} \sum_{k=0}^{\infty} a_k(\xi - x_0)^k\,d\xi = \int_{x_0}^{x} \lim_{n\to\infty} \sum_{k=0}^{n} a_k(\xi - x_0)^k\,d\xi$$

$$= \lim_{n\to\infty} \int_{x_0}^{x} \sum_{k=0}^{n} a_k(\xi - x_0)^k\,d\xi = \lim_{n\to\infty} \sum_{k=0}^{n} \frac{a_k}{k+1}(x-x_0)^{k+1} = \sum_{k=0}^{\infty} \frac{a_k}{k+1}(x-x_0)^{k+1}.$$

The equality of the first and second lines in the above equation (interchanging the limit and the integral) is guaranteed by the uniform convergence theorem, Theorem 4.2.2.

For the differentiation result, we need to first show that the power series generated by term-by-term differentiation is uniformly convergent on the interval $[x_0 - R_1, x_0 + R_1]$ for any $0 < R_1 < R$. Notice that if

$$\lim_{n\to\infty} \frac{|a_{n+1}|}{|a_n|} = L,$$

then

$$\lim_{k\to\infty} \frac{|(k+1)a_k|}{|ka_{k-1}|} = \left(\lim_{k\to\infty} \frac{k+1}{k}\right)\left(\lim_{k\to\infty} \frac{|a_k|}{|a_{k-1}|}\right) = 1 \cdot L = L.$$

So provided that the limit in Theorem 4.3.1 exists, the power series in the differentiation case converges absolutely with the same radius of convergence, R, as the original series and converges uniformly on $[x_0 - R_1, x_0 + R_1]\ \forall R_1 < R$. A somewhat more complicated argument similar to the one in Exercise B.4.9 works when the limit in Theorem 4.3.1 does not exist.

Now by Theorem 4.2.1, since the series

$$\sum_{k=1}^{\infty} ka_k(x-x_0)^{k-1}$$

converges uniformly on $[x_0 - R_1, x_0 + R_1]$ for any $0 < R_1 < R$ and its partial sums are continuous polynomials, it must be continuous as a function of x on any of these intervals. Since R_1 can again be chosen arbitrarily close to R, this series must be continuous on $(x_0 - R, x_0 + R)$, and therefore it must be integrable on any closed subinterval in $(x_0 - R, x_0 + R)$. So again the uniform convergence theorem can be applied to obtain the original power series as the antiderivative that is zero at $x = x_0$. By the fundamental theorem of calculus, the derivative of this antiderivative (that is, the derivative of the original power series) must be the desired power series which is the term-by-term derivative of the original power series. ☐

4.5 Taylor series

Until now, we have only obtained the power series for one function:

$$f(x) = \frac{1}{1-x} = \sum_{n=0}^{\infty} x^n$$

for $|x| < 1$. This power series can be recognized, of course, because it is a geometric series. In general, how can the coefficients a_n be chosen so that a power series represents a certain specific function? The answer to this question was given in 1715 by Brook Taylor.[3]

To understand this relationship, consider a function defined as a power series

$$f(x) := \sum_{n=0}^{\infty} a_n (x - x_0)^n$$

having a positive radius of convergence $R > 0$. From Theorem 4.4.1, this power series can be differentiated term-by-term, yielding

$$f'(x) = \sum_{n=1}^{\infty} n a_n (x - x_0)^{n-1}.$$

But then notice what happens at $x = x_0$:

$$f'(x_0) = \sum_{n=1}^{\infty} n a_n (x_0 - x_0)^{n-1} = a_1,$$

since every term in the series is zero accept for the first term. (Notice that in series notation, 0^0 must be understood as 1.) Similarly, taking the second derivative yields

$$f''(x_0) = \sum_{n=2}^{\infty} n(n-1) a_n (x_0 - x_0)^{n-2} = 2a_2.$$

---

3 Brook Taylor (1685–1731) was an English mathematician who introduced Taylor series and the Taylor theorem in 1715, though their significance was not understood until many years later.

Continuing this process, one finds that in general a_n is given in terms of the n-th derivative evaluated at the center:

$$f^{(n)}(x_0) = n!a_n \quad \Longleftrightarrow \quad a_n = \frac{f^{(n)}(x_0)}{n!},$$

or going back to the original series,

$$f(x) = \sum_{n=0}^{\infty} \frac{f^{(n)}(x_0)}{n!}(x - x_0)^n.$$

The above series is the **Taylor series** for the function f. Any function that can be represented as a power series in terms of its derivatives in this way having a positive radius of convergence is called an **analytic function**. When the center of the interval of convergence is $x_0 = 0$, the series can be called a **Maclaurin**[4] **series**. One of the most important implications for Taylor series is that if we know all of the derivatives of a function, then we know at least one way to represent it as a power series.

The following three examples show how to use the definition to find the Taylor series by direct computation for functions whose derivatives form patterns, at least when evaluated at some convenient center, x_0.

Example 4.5.1. Please find the Taylor series for the function $f(x) = e^x$ centered at $x_0 = 0$ (that is, the Maclaurin series) and find the radius of convergence, R.

Answer. This function is the simplest to start with because it has perhaps the easiest representation for its derivatives:

$$f(x) = e^x \quad \Longrightarrow \quad f^{(n)}(x) = e^x \quad \Longrightarrow \quad f^{(n)}(0) = 1.$$

Indeed, the exponential function is essentially the only function whose derivative is itself. Therefore, the Taylor series for this function is straightforward:

$$f(x) = e^x = \sum_{n=0}^{\infty} \frac{1}{n!}x^n = \sum_{n=0}^{\infty} \frac{x^n}{n!}.$$

Now to find the radius of convergence, compute the limit using Theorem 4.3.1:

$$L = \lim_{n\to\infty} \frac{1/(n+1)!}{1/n!} = \lim_{n\to\infty} \frac{n!}{(n+1)!} = \lim_{n\to\infty} \frac{1}{n+1} = 0.$$

Hence, the radius of convergence is $R = +\infty$, implying that this Taylor series converges for all real values of x.

---

4 Colin Maclaurin (1698–1746) was a Scottish mathematician and early supporter of Isaac Newton.

Example 4.5.2. Please find the Maclaurin series (that is, the Taylor series centered at $x_0 = 0$) for the function $f(x) = \sin x$ and find the radius of convergence, R.

Answer. This time, the pattern for derivatives is slightly more complicated:

$$f(x) = \sin x \implies f'(x) = \cos x \implies f''(x) = -\sin x \implies f'''(x) = -\cos x.$$

After the third derivative, all of the derivatives follow the same pattern,

$$f^{(4)}(x) = \sin x \implies f^{(5)}(x) = \cos x \implies f^{(6)}(x) = -\sin x$$
$$\implies f^{(7)}(x) = -\cos x,$$

and in general,

$$f^{(4k)}(x) = \sin x \implies f^{(4k+1)}(x) = \cos x \implies f^{(4k+2)}(x) = -\sin x$$
$$\implies f^{(4k+3)}(x) = -\cos x,$$

where $k \in \mathbb{Z}^+$. Now when these derivatives are evaluated at the center $x_0 = 0$, all of the even derivatives are zero:

$$f^{(4k)}(0) = 0, \quad f^{(4k+1)}(0) = 1, \quad f^{(4k+2)}(0) = 0, \quad f^{(4k+3)}(0) = -1.$$

Therefore, in the Maclaurin series for the sine function, all of the even terms vanish and only the odd terms are left, with an alternating sign:

$$f(x) = \sin x = \sum_{k=0}^{\infty} \frac{(-1)^k x^{2k+1}}{(2k+1)!}.$$

Now the limit in Theorem 4.3.1 yields the radius of convergence:

$$L = \lim_{k\to\infty} \frac{|(-1)^{2k+3}/(2k+3)!|}{|(-1)^{2k+1}/(2k+1)!|} = \lim_{k\to\infty} \frac{(2k+1)!}{(2k+3)!} = \lim_{k\to\infty} \frac{1}{(2k+2)(2k+3)} = 0.$$

Hence the radius of convergence is again $R = +\infty$, implying that this Maclaurin series converges for all real values of x. In addition, this series makes clear that the sine function has odd symmetry about the y-axis since this series has only odd powers of x.

Example 4.5.3. Please find the Taylor series for the function $f(x) = 1/x$ centered at $x_0 = 1$ and find the radius of convergence, R.

Answer. Notice that in this case, neither the function nor any of its derivatives can be evaluated at $x_0 = 0$ since this x value is not in its domain. But its derivatives do follow a simple pattern, and these can be evaluated at $x_0 = 1$:

$$f(x) = \frac{1}{x} \implies f^{(n)}(x) = \frac{(-1)^n n!}{x^{n+1}} \implies f^{(n)}(1) = (-1)^n n!$$

So the Taylor series for the reciprocal function is just

$$f(x) = 1/x = \sum_{n=0}^{\infty} (-1)^n (x-1)^n.$$

Again to find the radius of convergence, compute the limit using Theorem 4.3.1:

$$L = \lim_{n \to \infty} \frac{|(-1)^{n+1}|}{|(-1)^n|} = 1.$$

Hence the radius of convergence is $R = 1$; notice that this is in fact the distance between the center x_0 and the singularity of the reciprocal function at $x = 0$.

There are other functions whose Taylor and Maclaurin series can be found by direct computation, and some of these are dealt with in the exercises. To really increase the range of functions whose Taylor and Maclaurin series are known, however, a new approach is needed, and this is discussed next.

4.5.1 Six standard series

So far, Taylor series for a number of functions have been derived by direct computation using their definition. Now Taylor series for several functions that cannot easily be found by direct computation are derived based on six standard Maclaurin series (Taylor series with $x_0 = 0$) which can be found either in the examples just above or in the exercises at the end of this chapter:

- $\dfrac{1}{1-x} = \sum_{k=0}^{\infty} x^k$ for $|x| < 1$,

- $\sin x = \sum_{k=0}^{\infty} \dfrac{(-1)^k x^{2k+1}}{(2k+1)!}$ for $x \in \mathbb{R}$,

- $\dfrac{1}{1+x} = \sum_{k=0}^{\infty} (-1)^k x^k$ for $|x| < 1$,

- $\cos x = \sum_{k=0}^{\infty} \dfrac{(-1)^k x^{2k}}{(2k)!}$ for $x \in \mathbb{R}$,

- $e^x = \sum_{k=0}^{\infty} \dfrac{x^k}{k!}$ for $x \in \mathbb{R}$,

- $\ln(x+1) = \sum_{k=1}^{\infty} \dfrac{(-1)^{k-1} x^k}{k}$ for $|x| < 1$.

Everyone who works with Taylor series should know these six Maclaurin series by heart because they can be used to find Taylor series for a wide variety of functions. Certain other series may be worth remembering too, but our discussion here will be limited to these six. NHL fans can think of these as the original six.

The next several examples demonstrate how one can use substitution, trigonometric identities, calculus and the six Maclaurin series listed above to derive Taylor or Maclaurin series for a variety of functions.

Example 4.5.4. Find the Maclaurin series for $\sin x^2$ and give the interval of convergence.

Answer. Computing this series directly from the definition would be difficult since the chain and product rules would generate more and more complicated derivatives. A much simpler approach for finding this series is to notice that

$$\sin u = \sum_{k=0}^{\infty} \frac{(-1)^k u^{2k+1}}{(2k+1)!}$$

for all $u \in \mathbb{R}$, so one can substitute $u = x^2$ into this expansion to obtain

$$\sin x^2 = \sum_{k=0}^{\infty} \frac{(-1)^k x^{4k+2}}{(2k+1)!},$$

and this series must converge for all $x \in \mathbb{R}$.

Example 4.5.5. Find a series expansion for $\exp(x + 1) \equiv e^{x+1}$ that is valid for all $x \in \mathbb{R}$.

Answer. Notice that because of the way that this question is posed, there are many possible answers. One immediate answer is to use the Maclaurin series for e^u and then substitute $u = x + 1$. The result is a Taylor series centered at $x_0 = -1$ with an infinity radius of convergence:

$$\exp(x + 1) = \sum_{k=0}^{\infty} \frac{(x + 1)^k}{k!}.$$

A second expansion follows from noting that $\exp(x + 1) = e^{x+1} = e\, e^x$, so

$$\exp(x + 1) = e \sum_{k=0}^{\infty} \frac{x^k}{k!} = \sum_{k=0}^{\infty} \frac{e x^k}{k!}$$

is its Maclaurin series, also valid for all values of x. Which of these series is better? The answer depends on what one wishes to use it for. In some cases, one might prefer the Maclaurin series, while in other cases, the Taylor series would be the better choice.

Example 4.5.6. Please find the Maclaurin series for the function $f(x) = 1/(7 + x)$ and find the radius of convergence, R.

Answer. Here f looks a bit like the function in Example 4.5.3, but also like the second entry in the first column of the six standard series. Which example should be followed? The key observation is that replacing x by $x + 7$ in the series in Example 4.5.3 will lead to a Taylor series centered at $x_0 = 7$, not a Maclaurin series. Notice, on the other hand, that

$$\frac{1}{7 + x} = \frac{1}{7}\left(\frac{1}{1 + (x/7)} \right),$$

so for the standard series for $1/(1 + u)$ with $u = x/7$ we have

$$f(x) = \frac{1}{7+x} = \frac{1}{7}\left(\frac{1}{1+u}\right) = \frac{1}{7}\sum_{k=0}^{\infty}(-1)^k u^k = \frac{1}{7}\sum_{k=0}^{\infty}\left(\frac{-1}{7}\right)^k x^k.$$

This final power series is the desired Maclaurin series.

Example 4.5.7. Please find the Maclaurin series expansion for $\sin^2\theta$.

Answer. One might be tempted find the Maclaurin series for $\sin\theta$ and multiply it by itself. While it is true that sometimes one must find a result by multiplying two series, in general this is not the preferred approach. In this case, it is best to be aware of the appropriate trigonometric identity, the half-angle/double-angle formula

$$\sin^2\theta = \frac{1-\cos 2\theta}{2} = \frac{1}{2} - \frac{1}{2}\cos 2\theta.$$

Using the Maclaurin series for $\cos u$ with $u = 2\theta$, one finds that

$$\sin^2\theta = \frac{1}{2} - \frac{1}{2}\sum_{k=0}^{\infty}\frac{(-1)^k(2\theta)^{2k}}{(2k)!} = \sum_{k=1}^{\infty}\frac{(-1)^k 2^{2k-1}}{(2k)!}\theta^{2k}.$$

Notice that the $k = 0$ term cancels with the one-half out front.

The final example of this set concerns using calculus to find a Taylor series for a given function by realizing that this function is the derivative or the integral of some other function whose Taylor series is known.

Example 4.5.8. Please find the Maclaurin series for

$$f(x) = \frac{1}{(1+x)^2}.$$

Answer. Notice that f looks very similar to the second of the standard six Maclaurin series, except that here the denominator is squared. In this case, the key observation is that f is essentially the derivative of the known Maclaurin series, and this leads us to its Maclaurin series:

$$f(x) = -\frac{d}{dx}\left(\frac{1}{1+x}\right)$$

$$= -\frac{d}{dx}\left(\sum_{k=0}^{\infty}(-1)^k x^k\right)$$

$$= -\sum_{k=1}^{\infty}k(-1)^k x^{k-1}$$

$$= \sum_{n=0}^{\infty}(n+1)(-1)^n x^n$$

Of course one could find this series by direct computation, but the approach above is easier, and there are other cases where the series is much harder to find if one does not notice that it is the derivative or the integral of something known (see Exercise A.4.3).

4.6 Taylor theorem

The previous section discussed the work of Brook Taylor regarding power series; the current section deals with how one can express the error that is made when one replaces a Taylor series by its partial sum—the Taylor polynomial.

Theorem 4.6.1 (Taylor). *Suppose that for some $n \in \mathbb{Z}^+$, the function f has $n+1$ derivatives on some interval $(x_0 - R, x_0 + R)$, where $x_0 \in \mathbb{R}$ and $R > 0$. Then $f(x) = P_n(x) + R_n(x)$, where*

$$P_n(x) := \sum_{k=0}^{n} \frac{f^{(k)}(x_0)}{k!}(x - x_0)^k$$

*is the n-th **Taylor polynomial** and*

$$R_n(x) := \frac{f^{(n+1)}(c)}{(n+1)!}(x - x_0)^{n+1},$$

*where c is between x_0 and x. The function R_n is the n-th **Taylor remainder**.*

Proof. The proof of this result is a relatively easy application of the mean value theorem; see Exercise B.4.10. □

Remark. In terms of approximations, Taylor's theorem is much more helpful than the Taylor series because of the Taylor remainder term. One can almost never know the exact value of the Taylor remainder, but frequently we can bound this remainder and make it small for all x in some interval. So we can know that $|f(x) - P_n(x)|$ is small on that interval, making P_n a good approximation of f.

The next several examples present applications of Taylor series and the Taylor theorem, particularly showing how they can be used to approximate functions or integrals that are difficult or impossible to compute by hand. In some ways, this entire text has been a buildup to these examples.

Example 4.6.1. Please compute the limit

$$\lim_{x \to 0} \frac{2e^{-x} - 2 + 2x - x^2}{x^3}$$

using Taylor series.

Answer. One could compute this limit by repeatedly applying l'Hôpital's rule, but notice that by cancelation:

$$\lim_{x\to 0}\frac{2e^{-x}-2+2x-x^2}{x^3} = \lim_{x\to 0}\frac{2(\sum_{k=0}^{\infty}\frac{(-x)^k}{k!})-2+2x-x^2}{x^3} = \lim_{x\to 0}\frac{2\sum_{k=3}^{\infty}\frac{(-x)^k}{k!}}{x^3}$$

$$= -2\lim_{x\to 0}\sum_{k=3}^{\infty}\frac{(-x)^{k-3}}{k!} = -2\lim_{x\to 0}\sum_{m=0}^{\infty}\frac{(-x)^m}{(m+3)!} = -1/3.$$

In the end, the limit is computed by simply picking off the $m = 0$ coefficient since all of the other terms go to zero as $x \to 0$.

Example 4.6.2. Use Taylor polynomials centered at $x_0 = 0$ to approximate $f(x) = \sin x$ on the interval $[-\pi, \pi]$ to an accuracy of 0.01.

Answer. Notice that $|f(x) - P_n(x)| = |R_n(x)|$, meaning that if for some n, the Taylor remainder is absolutely bounded by 0.01, then the Taylor polynomial for that n will give the desired approximation. In this case,

$$|R_n(x)| = \left|\frac{f^{(n+1)}(c)}{(n+1)!}x^{n+1}\right| \le \frac{1}{(n+1)!}\pi^{n+1}$$

because $x \in [-\pi, \pi]$ and $|f^{(n+1)}(c)| \le 1$ (both $\sin x$ and $\cos x$ are absolutely bounded by $1\ \forall x \in [-\pi, \pi]$). So $|R_1(x)| \le \pi^2/2 \approx 4.9348$, $|R_2(x)| \le \pi^3/6 \approx 5.1677$, and so forth, which does not seem like a good start. Skipping ahead a bit, though, one finds that $|R_8(x)| \le \pi^9/9! \approx 0.0821$, $|R_9(x)| \le \pi^{10}/10! \approx 0.0258$, and $|R_{10}(x)| \le \pi^{11}/11! \approx 0.0074$. Thus, the tenth remainder is small enough, and the desired polynomial is $P_9(x) = P_{10}(x)$ (these are the same because the coefficients of the even powers are all zero in this expansion):

$$P_9(x) = P_{10}(x) = x - \frac{x^3}{6} + \frac{x^5}{120} - \frac{x^7}{5040} + \frac{x^9}{362880}.$$

It likely is difficult if not impossible to get a sense of why this is a good approximation by just looking at this polynomial. A much better sense can be obtained by looking at a plot (produced for example by a graphics calculator or a computer algebra system [CAS]), such as the one in Figure 4.7.

Example 4.6.3. Compute to four decimal places

$$\int_0^1 e^{-x^2}\,dx.$$

Answer. Since there is no closed-form antiderivative for the integrand, one must find an approximation for e^{-x^2}. Recall that the formal Taylor series is

$$e^u = \sum_{k=0}^{+\infty}\frac{u^k}{k!},$$

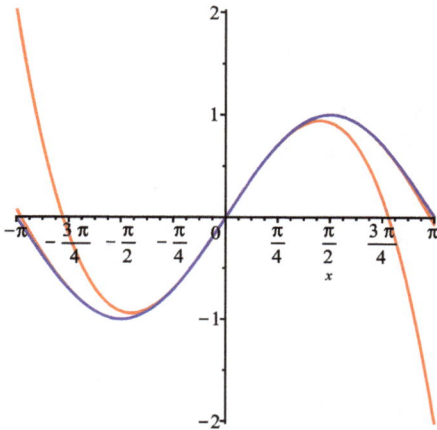

Figure 4.7: Sine approximation on $[-\pi, \pi]$. The red curves are the Taylor polynomials $P_3(x)$ and $P_7(x)$; the blue curve is $P_9(x)$, which on this scale is indistinguishable from a plot of $\sin x$. Notice that the worst convergence is always near the outer edge of the interval, far from the center $x = 0$. Which red curve is $P_3(x)$ and which is $P_7(x)$?

so by the Taylor theorem,

$$e^u = \sum_{k=0}^{n} \frac{u^k}{k!} + \frac{e^c}{(n+1)!} u^{n+1}$$

for any $n \in \mathbb{Z}^+$, where c is some number between 0 and u. The first term in this sum is the Taylor polynomial $P_n(u)$, while the second term is the Taylor remainder $R_{n+1}(u)$. To apply this result to the problem at hand, let $u = -x^2$; then $u \in [-1, 0]$ since $x \in [0, 1]$ for this integral. Thus, $c \in (-1, 0)$ and $e^c < e^0 = 1$, so

$$\left| \frac{e^c}{(n+1)!} (-x^2)^{n+1} \right| < \frac{x^{2n+2}}{(n+1)!}.$$

Integrating this bound on the error term, one obtains a bound on the error term for the integral:

$$\left| \int_0^1 R_{n+1}(-x^2) \, dx \right| < \frac{1}{(n+1)!(2n+3)}.$$

When $n = 5$, this bound is less than 0.00011, which is slightly too large; when $n = 6$, the bound is less than 0.00002, which is likely small enough.

Now integrating the polynomial, one finds

$$\int_0^1 e^{-x^2} \, dx \approx \int_0^1 P_6(-x^2) \, dx$$

$$= \int_0^1 \left(1 - x^2 + \frac{x^4}{2} - \frac{x^6}{6} + \frac{x^8}{24} - \frac{x^{10}}{120} + \frac{x^{12}}{720}\right) dx$$

$$= 1 - \frac{1}{3} + \frac{1}{10} - \frac{1}{42} + \frac{1}{216} - \frac{1}{1320} + \frac{1}{9360}$$

$$\approx 0.746836,$$

so the four-decimal-place approximation is 0.7468.

4.7 Limitations of analyticity

Until now, this chapter has discussed Taylor series, the Taylor theorem, power series, and analytic functions.[5] It may seem that we can find a power series for any function, and indeed there was a time when many mathematicians felt that any significant function could be expressed and studied as a power series. Some even felt that f being a function should imply that f has a power series expansion. In the past 50 to 100 years, however, the importance of functions that definitely are not analytic has become clear in some settings. Indeed, the modern mathematical understanding of shock waves, for example, hinges on *test functions* that are infinitely differentiable, but not analytic.

The next two examples present two nonanalytic functions. The first is an example of a function that is differentiable, but the derivative is *not* continuous, and therefore not itself differentiable. The second is a function that is infinitely differentiable, but still not analytic because the radius of convergence would be zero.

Example 4.7.1. Give an example of a function f that *is* differentiable at a point, but where f' *is not* continuous at this point.

Answer. Let

$$f(x) = \begin{cases} x^2 \sin \frac{1}{x} & x \neq 0 \\ 0 & x = 0. \end{cases}$$

Then for $x \neq 0$, the derivative is relatively easy to compute using the standard rules of calculus. At $x = 0$, however, the standard rules of calculus do not apply, and one must use the definition of the derivative:

$$f'(0) := \lim_{x \to 0} \frac{f(x) - f(0)}{x - 0} = \lim_{x \to 0} x \sin \frac{1}{x} = 0.$$

Combining all this, one finds

---

[5] Recall that a function is *analytic* iff it has a convergent Taylor series having a positive radius of convergence.

$$f'(x) = \begin{cases} 2x\sin\frac{1}{x} - \cos\frac{1}{x} & x \neq 0 \\ 0 & x = 0. \end{cases}$$

But notice that while $f'(0) = 0$,

$$\lim_{x\to 0} f'(x) \quad \text{DNE}$$

because $-\cos(1/x)$ oscillates wildly between -1 and 1 as x approaches zero. So f' is not continuous at $x = 0$, and hence $f''(0)$ does not exist.

Example 4.7.2. Give an example of a function f that is infinitely differentiable at $x = 0$, but whose Maclaurin series has a radius of convergence $R = 0$.

Answer. Consider the function

$$f(x) = \begin{cases} e^{-1/x^2} & x > 0 \\ 0 & x \leq 0. \end{cases}$$

Again, for $x \neq 0$, the derivative is relatively easy to compute using the standard rules of calculus:

$$f'(x) = \begin{cases} \frac{2}{x^3} e^{-1/x^2} & x > 0 \\ 0 & x < 0. \end{cases}$$

And again at $x = 0$, one must use the definition of the derivative:

$$f'(0) := \lim_{x\to 0} \frac{f(x) - f(0)}{x - 0} = \lim_{x\to 0} \frac{e^{-1/x^2}}{x} = \lim_{x\to 0} \frac{1/x}{e^{1/x^2}} = \lim_{x\to 0} \frac{x}{2e^{1/x^2}} = 0.$$

The second to last equality above is obtained by the l'Hôpital rule. Hence in full,

$$f'(x) = \begin{cases} \frac{2}{x^3} e^{-1/x^2} & x > 0 \\ 0 & x \leq 0. \end{cases}$$

One can continue to compute the higher derivatives of this function f, although they do become increasingly messy. Exercise B.4.11 asks for the second derivative $f''(x)$. In general, the form of the n-th derivative is

$$f^{(n)}(x) = \begin{cases} Q_n(x) e^{-1/x^2} & x > 0 \\ 0 & x \leq 0, \end{cases}$$

where $Q_n(x)$ is some rational expression that becomes increasingly complicated as n increases. But no matter how complicated this expression is, the l'Hôpital rule always yields that the exponential wins and that the next derivative $f^{(n+1)}(0) = 0 \; \forall n \in \mathbb{Z}^+$.

Now recall, that for any function f, the Maclaurin series is

$$\sum_{k=0}^{\infty} \frac{f^{(k)}(0)}{k!} x^k.$$

What is the Maclaurin series for the function f in this example? Since $f^{(k)}(0) = 0 \ \forall \ k \in \mathbb{N}$, the entire Maclaurin series collapses to nothing:

$$\sum_{k=0}^{\infty} \frac{f^{(k)}(0)}{k!} x^k \equiv 0$$

For $x > 0$, this function is e^{-1/x^2}, which is always positive. Hence the radius of convergence is just $R = 0$. Therefore in this case, a Taylor polynomial or Taylor series approximation is not possible.

Exercises 4, Group A

4.1. Please determine the interval of convergence for each of the following power series. There is no need to check the endpoints. Justify your answer.

(a) $\sum_{n=1}^{\infty} \frac{x^n}{n}$

(b) $\sum_{k=0}^{\infty} \frac{x^k}{k!}$

(c) $\sum_{j=0}^{\infty} (x/5)^j$

(d) $\sum_{m=1}^{\infty} \frac{(-1)^m x^m}{3m}$

(e) $\sum_{k=0}^{\infty} \frac{(7x-4)^k}{2^k}$

(f) $\sum_{n=1}^{\infty} \frac{n(x+2)^n}{5^n}$

(g) $\sum_{j=1}^{\infty} \frac{x^j}{\sqrt{2j}}$

(h) $\sum_{k=0}^{\infty} (-ex/\pi)^k$

(i) $\sum_{n=1}^{\infty} nx^n$

(j) $\sum_{k=0}^{\infty} k!x^k$

(k) $\sum_{m=2}^{\infty} \frac{x^m}{m \ln m}$

(l) $\sum_{n=1}^{\infty} \frac{2^n - 3^n}{5^n} x^n$

(m) $\sum_{m=1}^{\infty} 2^m (7x-4)^m$

(n) $\sum_{k=1}^{\infty} \frac{(\ln k^k) x^k}{5^k}$

(o) $\sum_{j=1}^{\infty} \frac{j^j}{j!} x^j$

(p) $\sum_{i=0}^{\infty} \frac{7^i - 5}{\sqrt{i!}} x^i$

Answers. (a) $|x| < 1$. (b) $(-\infty, \infty)$. (j) $\{0\}$. (l) $|x| < 5/3$. (m) $(1/2, 9/14)$. (n) $|x| < 5$.

4.2. Please find a power series for each of the following integrals by writing the integrand as a power series and then integrating the result.

(a) $\displaystyle\int_0^x \frac{\sin t}{t}\,dt$

(e) $\displaystyle\int_0^x \frac{t^3}{1+t}\,dt,\ |x| < 1$

(b) $\displaystyle\int_0^x \frac{1 - e^{-t}}{t}\,dt$

(f) $\displaystyle\int_0^x \frac{1 - e^{-t}}{t}\,dt$

(c) $\tanh^{-1} x := \displaystyle\int_0^x \frac{dt}{1 - t^2},\ |x| < 1$

(g) $\displaystyle\int_1^x \frac{\cos t}{t^2}\,dt,\ x > 1$

(d) $\displaystyle\int_0^x \frac{\cos t - 1}{t}\,dt$

(h) $\displaystyle\int_0^x \frac{\cos t - 1}{t}\,dt$

Answers. (d) $\sum_{k=1}^\infty \frac{(-1)^k x^{2k}}{(2k)(2k)!}$. (e) $\sum_{k=0}^\infty \frac{(-1)^k x^{k+4}}{k+4}$.

4.3. Find the Maclaurin series for $f(x) = x^5 e^{5x}$ from the standard Maclaurin series for e^x. Then find the Maclaurin series for $g(x) = 5(x^4 + x^5)e^{5x}$.
   Hint: How are f and g related?

4.4. For each of the functions given below, use the six standard power series. Give the interval of convergence for each series. α and ω are constants.

(a) $f(x) = \dfrac{x^2}{1+x}$

(g) $h(z) = \dfrac{3}{7+z}$

(b) $g(x) = \sin(2x^2)$

(h) $f(x) = \dfrac{x+1}{x^2-1}$

(c) $h(x) = \ln(2+x)$ (center?)

(i) $h(x) = \dfrac{\ln(2+x)}{x}$

(d) $f(x) = \dfrac{x}{1-x^2}$

(j) $f(x) = \dfrac{1-x}{e^x}$

(e) $h(x) = e^{x^2+1}$

(k) $g(t) = \sin(2t+1)$ (center?)

(f) $g(x) = \dfrac{x^2+1}{x^2-1}$

(l) $g(t) = e^{\alpha t}\cos\omega t$ (find the first five terms)

Answers. Hints. (a) $f(x) = \sum_{k=0}^\infty (-1)^k x^{k+2}$ for $|x| < 1$. (c) $h(x) = \sum_{k=1}^\infty \frac{(-1)^{k-1}(x+1)^k}{k}$ for $|x+1| < 1$. Center is $x = -1$. (g) **Hint:** $h(z) = \frac{3}{7}\frac{1}{1+z/7}$. (k) Center is $t = -1/2$.

Exercises 4, Group B

4.1. Compute the Maclaurin series (Taylor series with $x_0 = 0$) for $f(x) = \ln(x+1)$ by directly computing the derivatives (the direct method).

4.2. Please show that

$$\sum_{n=1}^{\infty} \frac{(-1)^n}{n + x^6}$$

is uniformly convergent $\forall x \in \mathbb{R}$, but not (pointwise) absolutely convergent for any $x \in \mathbb{R}$.

Hint: Notice that this is **not** a power series, so the Taylor theorem does **not** apply. It is, however, an alternating series, so the difference between the series and $S_n(x)$ (the n-th partial sum) is bounded by the next term in the series, $a_{n+1} = 1/(n + x^6)$.

4.3. Suppose $f_n : [0, \pi] \to \mathbb{R}$ by $f_n(x) = \frac{\sin nx}{n}$.
(a) This sequence of functions converges pointwise to a limit as $n \to \infty$; what is the limit function f?
(b) Does this sequence of functions converge uniformly? Please explain why or why not.
    Hint: Find a uniform bound independent of x.

4.4. Plot f_1, f_3, f_5 and f_9 for the sequence of functions in Example 4.1.3,

$$f_n(x) = \frac{nx}{1 + n^2 x^2},$$

defined on the domain $\mathcal{D} = [0, 1]$.

4.5. Suppose $f_n : [0, 1] \to \mathbb{R}$ by $f_n(x) = nxe^{-nx^2}$.
(a) Does this sequence of functions converge pointwise to a limit as $n \to \infty$? If so, what is the limit function f?
(b) Does this sequence of functions converge uniformly? Please explain why or why not.
    Hint: Find the location of the maximum value of $f_n(x)$ on $[0, 1]$, and then find the value of f_n at that maximum.

4.6. Consider the sequence of functions

$$\left\{ \sqrt{x}, \sqrt{x + \sqrt{x}}, \sqrt{x + \sqrt{x + \sqrt{x}}}, \ldots \right\},$$

each defined on the interval $[0, 1]$. Use a variation on the argument in Example 2.2.2 to show that this sequence of functions converges pointwise on $[0, 1]$, but notice that the resulting limit function is *not* continuous, so convergence here cannot be uniform.

4.7. Please complete the proof of Theorem 4.3.1 by showing that convergence is uniform on the interval $[-R_1, R_1]$ for any $R_1 < R$.
    Hint: Show that the difference between the power series and its partial sum

$$\left| \sum_{k=0}^{n} a_k x^k - \sum_{k=0}^{\infty} a_k x^k \right|$$

can be uniformly bounded by the appropriate convergent series that is independent of x and tends to zero as n increases.

4.8. Consider the power series

$$\sum_{n=0}^{\infty} a_n x^n,$$

where

$$a_n = \begin{cases} 1 & n \text{ is odd} \\ \frac{1}{n} & n \text{ is even.} \end{cases}$$

Show that this power series converges for $x \in (-1, 1)$ even though

$$\lim_{n \to \infty} \frac{|a_{n+1}|}{|a_n|}$$

does not exist.

Hint: Consider the comparison test and appropriately chosen geometric series.

4.9. Consider the power series

$$\sum_{n=0}^{\infty} a_n x^n$$

and suppose the limit in Theorem 4.3.1 does not exist as a finite real and is not infinity. Please show that nonetheless, there is still an interval of convergence as in Figure 4.5.

Hint: Pick a $c > 0$; notice that the power series

$$\sum_{n=0}^{\infty} a_n c^n$$

either converges or diverges. If it converges, explain why the original series converges for $|x| < c$, and if it diverges, explain why the original series diverges for $|x| > c$.

4.10. Please prove the Taylor theorem, Theorem 4.6.1.

Hint: Apply the mean value theorem to

$$F(x) = \sum_{k=0}^{n} \frac{f^{(k)}(x)}{k!} (y - x)^k + \frac{R_n(y)}{(y - x_0)} (y - x)^{n+1}$$

by noting that $F(y) = f(y) = F(x_0)$ and thus $F'(c) = 0$ for some c between y and x_0.

4.11. For the function $f(x)$ in Example 4.7.2, use $f'(x)$ as computed in the example to compute $f''(x)$.

4.12. Please compute

$$\int_0^1 \frac{\sin x}{x}\,dx$$

to *five*-decimal-place accuracy.

4.13. Find the Taylor polynomial $P_n(x)$ centered at $x_0 = 0$ that approximates $f(x) = \cos x$ on the interval $[-\pi/2, \pi/2]$ to an accuracy of 0.001.

4.14. Find the Taylor polynomial $P_n(x)$ centered at $x_0 = 0$ that approximates $f(x) = e^x$ on the interval $[-1,1]$ to an accuracy of 0.001.

4.15. Show that $e \notin \mathbb{Q}^+$. That is, show that e is irrational.

   Hint: On the contrary, suppose that $e = p/q$ where $p, q \in \mathbb{Z}^+$. Because of the Taylor series, $e = e^1 = \sum_{k=0}^\infty 1/(k!)$, meaning that

$$p(q-1)! = q!e = q! \sum_{k=0}^q \frac{1}{k!} + q! \sum_{k=q+1}^\infty \frac{1}{k!}$$

(please explain why all the above is true). Now $p(q-1)! - q! \sum_{k=0}^q 1/k!$ is a positive integer (why?), while

$$q! \sum_{k=q+1}^\infty \frac{1}{k!} < \sum_{n=1}^\infty \left(\frac{1}{q+1}\right)^k = \frac{1}{q} \le 1$$

(again explain why). So there needs to be an integer between 0 and 1, and there are precious few of those.

5 Improper Riemann integrals

This final chapter deals with a topic that was already introduced during our discussion of the integral test. The present presentation is independent of that discussion, though if the reader recalls what was said there, that may help to understand the current material.

Improper Riemann integrals or just improper integrals are an extension of the concept of the definite Riemann integral in the same way that a series is an extension of the concept of the finite sum. There are two kinds of improper integrals. These are discussed in the two sections below.

5.1 Improper integrals of the first kind

If someone reads the fine print in the standard definition of the definite Riemann integral, there are two (sometimes overlooked) requirements: (1) the domain of integration is a finite interval, often $[a, b]$ for some $a, b \in \mathbb{R}$, and (2) f is a bounded, real-valued function defined on $[a, b]$. In defining an improper integral of the first kind, this first requirement is relaxed. The upper limit of integration is taken to be $+\infty$, or the lower limit of integration is taken to be $-\infty$, or both limits are infinite at the same time.

Definition. Suppose that $a \in \mathbb{R}$ is fixed and that f is a continuous, real-valued function defined on the semi-infinite interval $[a, +\infty)$. Then the **improper Riemann integral of the first kind** (or just the **improper integral**) of f on $[a, +\infty)$ is

$$\int_a^{+\infty} f(x)\, dx := \lim_{b \to +\infty} \int_a^b f(x)\, dx$$

provided that this limit exists *as a finite real number*. Similarly, if $b \in \mathbb{R}$ is fixed, and f is a continuous, real-valued function defined on the semi-infinite interval $(-\infty, b]$, then

$$\int_{-\infty}^b f(x)\, dx := \lim_{a \to -\infty} \int_a^b f(x)\, dx$$

provided that this limit exists *as a finite real number*. Finally,

$$\int_{-\infty}^{+\infty} f(x)\, dx := \int_{-\infty}^0 f(x)\, dx + \int_0^{+\infty} f(x)\, dx$$

provided that both these improper integrals exist. Whenever the limit in the definition of an improper integral does not exist, the improper integral is said to diverge and/or to not exist.

https://doi.org/10.1515/9783110768398-005

Remark. The assumption that f is continuous in our definition is not strictly required; it simply makes things easier since it guarantees that the definite integral of f on $[a, b]$ will exist for all $b > a$.

Example 5.1.1. For which values of $q \in \mathbb{R}$ is

$$\int_1^\infty x^q \, dx$$

a finite real number?

Answer. The convergence of integrals of this sort was already considered in Example 3.3.5, Example 3.3.6 and material following them. In this case, the definite integral inside the limit is not difficult to evaluate, at least if one is not bothered by the parameter q. For $q \neq -1$,

$$\int_1^\infty x^q \, dx := \lim_{b \to \infty} \int_1^b x^q \, dx = \lim_{b \to \infty} \left(\frac{x^{q+1}}{q+1} \Big|_1^b \right) = \lim_{b \to \infty} \left(\frac{b^{q+1}}{q+1} - \frac{1}{q+1} \right).$$

The value of this integral depends on whether $q < -1$ or $q > -1$. When $q < -1$, b appears to a negative power in the limit, and hence the first term goes to zero and

$$\int_1^\infty x^q \, dx = \frac{-1}{q+1} = \frac{1}{|q+1|}.$$

When $q > -1$, b appears to a positive power in the limit, and hence the limit diverges to infinity, so the improper integral does not exist in this case.

What happens when $q = 1$? Then a new definition is needed (if one does not already know the definition of logarithm).

Definition. For $x > 0$, the **natural logarithm function** (or just the **logarithm**) is

$$\ln x := \int_1^x \frac{dt}{t}.$$

Notice that for every finite $x \in \mathbb{R}$, the integral in this definition must exist since the integrand $1/x$ is continuous for $x > 0$.

Notice that the integral test and the divergence of the harmonic series were both proven without considering the integral in the definition of the natural logarithm function. So the integral test and the divergence of the harmonic series together imply that

$$\lim_{b \to \infty} \ln b = +\infty$$

and the improper integral in this example does not exist when $q = -1$.

5.1.1 Natural logarithm

It is worth a brief aside to observe the basic properties of the natural logarithm function that follow more or less immediately from its definition.
- Notice that from the definition and the fundamental theorem of calculus,

$$\frac{d}{dx}(\ln x) \equiv \ln'(x) = \frac{1}{x} > 0$$

$\forall x > 0$. This means that the logarithm is a strictly increasing function on its domain. Indeed, as was discussed above, the logarithm grows (if very slowly) without bound as x increases. A plot of $y = \ln x$ is shown in Figure 5.1.

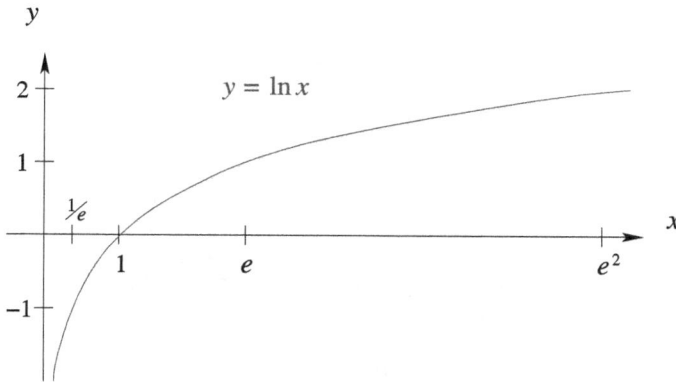

Figure 5.1: Plot of $y = \ln x$ for $x \in (0, e^2]$. Notice that $\ln 1 = 0$, $\ln e = 1$, $\ln(1/e) = -1$, $\ln e^2 = 2$, and in general, $\ln e^n = n$.

- Again directly from the definition, $\ln 1 = 0$, $\ln x > 0$ for $x > 1$, and $\ln x < 0$ for $0 < x < 1$. As was observed above,

$$\lim_{x \to \infty} \ln x = +\infty$$

based on the integral test and the harmonic series.
- For $x, y > 0$, several basic properties of the logarithm function follow from a change of variables (see Exercise B.5.2 below):
  - $\ln(xy) = \ln x + \ln y$,
  - $\ln(x/y) = \ln x - \ln y$,

- $\ln(1/x) = -\ln x,$
- $\ln x \to -\infty$ as $x \to 0^+$.

5.1.2 Area representation

What does an improper integral represent? Recall that the simplest representation for the standard (proper) integral is as the area under the curve $y = f(x)$ from a to b when f is nonnegative and continuous (see Figure 5.2(a)). The same sort of representation is used for improper integrals: when f is nonnegative and continuous, it is the area under the curve $y = f(x)$ from a to ∞ (see Figure 5.2(b)). This works even though one of the dimensions of the region under the curve is not finite.

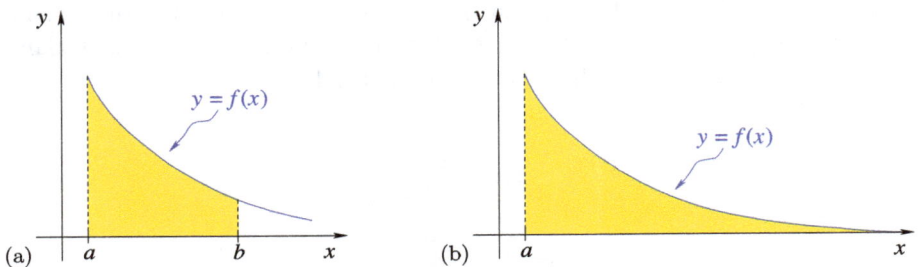

Figure 5.2: (a) The integral $\int_a^b f(x)\,dx$ as the area (in gold) under the curve $y = f(x)$ between a and b. (b) The improper integral $\int_a^\infty f(x)\,dx$ as the area (in gold) under the curve $y = f(x)$ from a onward.

This finite-area, infinite-length paradox has a famous generalization: Gabriel's horn (Gabriel's trumpet),[1] which has finite volume but infinite surface area.

Example 5.1.2. Consider the solid of revolution obtained by revolving the curve $y = f(x) = 1/x$ about the x-axis for $x \in [a, \infty)$ where $a > 0$ (see Figure 5.3). This object is often called Gabriel's horn. Find the volume inside Gabriel's horn, but also show that its length and (outer) surface area are infinite.

Answer. The horn extends from $x = a$ to $x = +\infty$, so its length is definitely infinite. To find the volume, slice the horn into infinitely thin disks having area equal to $\pi r^2 = \pi/x^2$ and thickness dx. So using this method of disks, one finds that the volume is given by an improper integral:

$$\int_a^{+\infty} \frac{\pi}{x^2}\,dx = \lim_{b \to +\infty} \int_a^b \frac{\pi}{x^2}\,dx = \lim_{b \to +\infty}\left(-\frac{\pi}{x}\Big|_a^b\right) = \lim_{b \to +\infty}\left(\frac{\pi}{a} - \frac{\pi}{b}\right) = \frac{\pi}{a}.$$

---

1 Gabriel is an archangel named in the Hebrew Bible (Old Testament), the New Testament and the Quran.

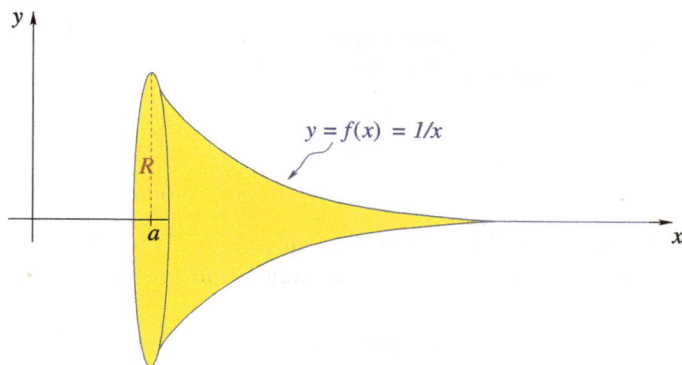

Figure 5.3: Gabriel's horn. The solid of revolution generated by spinning the curve $y = 1/x$ about the x-axis for $x \in [a, +\infty)$. Here the radius of the bell of the horn is $R = 1/a$.

For the surface area, the method of rings can be used where the thickness of each ring is again dx and the circumference is $2\pi r = 2\pi/x$. The surface area is then given by a slightly different improper integral:

$$\int_a^{+\infty} \frac{2\pi}{x}\, dx = \lim_{b \to +\infty} \int_a^b \frac{2\pi}{x}\, dx = \lim_{b \to +\infty} (2\pi \ln(x)|_a^b = \lim_{b \to +\infty} 2\pi(\ln(b) - \ln(a)) = +\infty.$$

That is, this integral diverges to infinity. Thus, the surface area is infinite, even though the volume is finite.

The next three examples help the reader understand two additional details that frequently arise when dealing with improper integrals of the first kind: integration by parts and doubly infinite integrals.

Example 5.1.3. Please compute

$$\int_1^\infty x e^{-x}\, dx.$$

Answer. Notice that the integrand here is the product of two factors: x, which is growing, and e^{-x}, which rapidly goes to zero. But does this second factor go to zero fast enough for the integral to converge? The answer can of course be found by applying the definition of the improper integral, but notice that the integral in this case requires integration by parts with $u = x$ and $dv = e^{-x}\, dx$:

$$\int_1^\infty x e^{-x}\, dx := \lim_{b \to \infty} \int_1^b x e^{-x}\, dx = \lim_{b \to \infty} \left(-x e^{-x}|_1^b + \int_1^b e^{-x}\, dx \right) = \lim_{b \to \infty} (e^{-1} - b e^{-b} - e^{-x}|_1^b) = \frac{2}{e}.$$

For readers familiar with integration by parts, the one-line calculation is likely sufficient. For those less familiar, the following subsection should help.

5.1.3 Integration by parts

Integration by parts is an integration technique based on the product rule. It is important to understand weak solutions to differential equations and is similar to results like Green's first identity.

Consider the product rule applied to two differentiable functions, u and v:

$$\frac{d}{dx}(u(x)v(x)) = u'(x)v(x) + u(x)v'(x).$$

Integrating both sides of this identity from a to b, one finds

$$\int_a^b \frac{d}{dx}(u(x)v(x))\,dx = \int_a^b (u'(x)v(x) + u(x)v'(x))\,dx.$$

But the integral on the left side above is the integral of a derivative, so it can be immediately evaluated (in other words, the integration and differentiation exactly cancel each other). So

$$(u(x)v(x))\big|_a^b = \int_a^b (u'(x)v(x) + u(x)v'(x))\,dx,$$

or

$$\int_a^b u(x)v'(x)\,dx = (u(x)v(x))\big|_a^b - \int_a^b u'(x)v(x)\,dx.$$

The previous formula is often remembered in its indefinite form:

$$\int u\,dv = uv - \int v\,du + c, \tag{5.1}$$

where c is an arbitrary constant.

Answer. Let us return to the solution for the previous example. The integrand that we wish to integrate, here $xe^x\,dx$, becomes $u\,dv$ in the integration by parts formula. For those not previously familiar with integration by parts, the most difficult choice that one must make in using this technique is deciding what to pick for u and what for dv. The two points that must be taken into account are (1) that whatever is chosen as dv must be (easily) integrable and (2) that what becomes $v\,du$ must also be integrable. In the case of the example at hand, notice that if $u = x$ is differentiated and $dv = e^{-x}\,dx$ is integrated (which is indeed not a difficult integral), then $v\,du = e^{-x}\,dx$, which can also be integrated.

Example 5.1.4. Please compute

$$\int_0^\infty x^3 e^{-x^2}\,dx.$$

Answer. This integral is somewhat similar to the one in Example 5.1.2, so it should be no surprise that integration by parts is again the main tool used in evaluating this integral. The main question then is what to make u and what dv. The typical expectation is that one should integrate the exponential and differentiate the power of x. But of course to easily integrate the exponential, there needs to be an $x\,dx$ paired with it. Thus, we let $u = x^2$ and $dv = e^{-x^2} x\,dx$. We have

$$\int_0^\infty x^3 e^{-x^2}\,dx := \lim_{b\to\infty} \int_0^b x^3 e^{-x^2}\,dx = \lim_{b\to\infty} \int_0^b (x^2)(e^{-x^2} x\,dx) = \lim_{b\to\infty} \left(\frac{-x^2}{2} e^{-x^2} \Big|_0^b + \int_0^b x e^{-x^2}\,dx \right).$$

The first term in the limit on the right above evaluates to zero, so

$$\int_0^\infty x^3 e^{-x^2}\,dx = \lim_{b\to\infty} \int_0^b x e^{-x^2}\,dx = \lim_{b\to\infty} \left(-\frac{1}{2} e^{-x^2} \Big|_0^b \right) = \frac{1}{2}.$$

Example 5.1.5. Please compute

$$\int_{-\infty}^\infty \frac{dx}{1 + x^2}.$$

Answer. The form of this integral is slightly different from those that have been considered so far. Notice that the lower limit of integration here is negative infinity. This doubly infinite integral requires that we break the integral at a convenient middle point into two integrals and then compute these two integrals separately. Zero is likely the most convenient middle point:

$$\int_{-\infty}^\infty \frac{dx}{1 + x^2} = \int_{-\infty}^0 \frac{dx}{1 + x^2} + \int_0^\infty \frac{dx}{1 + x^2} := \lim_{a\to-\infty} \int_a^0 \frac{dx}{1 + x^2} + \lim_{b\to\infty} \int_0^b \frac{dx}{1 + x^2}.$$

Now of course the integrals must be evaluated. Those who have seen this sort of integral before may recognize it as the arctan or $\tan^{-1}$ function. For those not familiar with this function, these integrals can be evaluated with the trigonometric substitution $x = \tan\theta \implies dx = \sec^2\theta\,d\theta$. Then for the first integral,

$$\lim_{a \to -\infty} \int_a^0 \frac{dx}{1+x^2} = \lim_{a \to -\infty} \int_{\tan^{-1}(a)}^{\tan^{-1}(0)} \frac{\sec^2 \theta}{\sec^2 \theta} \, d\theta = \lim_{a \to -\infty} \theta \Big|_{\tan^{-1}(a)}^0 = -\lim_{a \to -\infty} \tan^{-1}(a) = \frac{\pi}{2}.$$

The second integral evaluates much as the first to $\pi/2$; this is not surprising since the integrand is symmetric about zero. Thus,

$$\int_0^\infty \frac{dx}{1+x^2} = \frac{\pi}{2},$$

and adding the two sections together,

$$\int_{-\infty}^\infty \frac{dx}{1+x^2} = \pi.$$

It may not be possible to evaluate a certain improper integral exactly, but it may still be possible to show that the improper integral converges. Sometimes this is by comparison to a series via the integral test, and sometimes it is by comparison to another integral—one that can be evaluated.

Example 5.1.6. Does the integral

$$\int_{-\infty}^\infty \frac{dx}{1+x^8}$$

converge or diverge?

Answer. Notice that this time the issue is not to find the value of the integral, but rather simply to show that this integral converges. The key then is to find an integral whose value is known to be finite to compare the current integral to. The obvious integral to compare to in this case is the one in the previous example. Notice that for $x > 1$,

$$1 + x^2 < 1 + x^8 \quad \Longrightarrow \quad \frac{1}{1+x^8} < \frac{1}{1+x^2},$$

meaning that the integral in this example should be broken into three pieces:

$$\int_{-\infty}^\infty \frac{dx}{1+x^8} = \int_{-\infty}^{-1} \frac{dx}{1+x^8} + \int_{-1}^1 \frac{dx}{1+x^8} + \int_1^\infty \frac{dx}{1+x^8}.$$

The third part can then be bounded by $\pi/2$ since

$$\int_1^\infty \frac{dx}{1+x^8} \leq \int_1^\infty \frac{dx}{1+x^2} \leq \int_0^\infty \frac{dx}{1+x^2} = \frac{\pi}{2}.$$

By symmetry, the first part must have the same finite value as the third, and the second piece must be finite because it is the integral of a continuous function between finite bounds. Since all three pieces are finite, the original integral must also be.

5.2 Improper integrals of the second kind

Improper integrals of the second kind cover a somewhat broader class of improper integrals than do those of the first kind. Improper integrals of the second kind are integrals over a finite interval, but now the function behaves badly at some point on that interval. The bad behavior typically involves becoming unbounded at either end point of the interval or at some interior point of the interval.

Definition. Suppose that $a, b \in \mathbb{R}$ and that f is a continuous, real-valued function defined on the interval $[a, b)$. Then the **improper Riemann integral of the second kind** (or just the **improper integral**) of f on $[a, b]$ is

$$\int_a^b f(x)\,dx := \lim_{\beta \to b^-} \int_a^\beta f(x)\,dx$$

provided that this limit exists *as a finite real number*. Similarly, if f is a continuous, real-valued function defined on the interval $(a, b]$, then the **improper integral** of f on $[a, b]$ is

$$\int_a^b f(x)\,dx := \lim_{\delta \to a^+} \int_\delta^b f(x)\,dx$$

provided that this limit exists *as a finite real number*. Also, if $\exists\, c \in [a, b]$ such that f is a continuous, real-valued function defined on the intervals $[a, c)$ and $(c, b]$, then the **improper integral** of f on $[a, b]$ is

$$\int_a^b f(x)\,dx := \lim_{\xi \to c^-} \int_a^\xi f(x)\,dx + \lim_{\xi \to c^+} \int_\xi^b f(x)\,dx$$

provided that both of these limits exist *as finite real numbers*. Again, whenever the limit in the definition of an improper integral does not exist, the improper integral is said to diverge and/or to not exist.

The definition above may look complicated, but some basic examples should be helpful.

Example 5.2.7. Please compute

$$\int_0^1 \frac{dx}{\sqrt{x}}.$$

Answer. At first glance, this may appear to be just a standard Riemann integral, but on closer inspection, one should notice that the integrand $f(x) = \frac{1}{\sqrt{x}}$ is undefined at $x = 0$ and unbounded near $x = 0$. Because of this bad behavior near $x = 0$, this integral must be viewed as an improper integral:

$$\int_0^1 \frac{dx}{\sqrt{x}} := \lim_{\delta \to 0^+} \int_\delta^1 \frac{dx}{\sqrt{x}} = \lim_{\delta \to 0^+} \int_\delta^1 x^{-1/2}\, dx = \lim_{\delta \to 0^+} 2x^{1/2}\Big|_\delta^1 = 2 \lim_{\delta \to 0^+} (1 - \sqrt{\delta}) = 2.$$

In the previous example, if one had completely ignored the fact the integral is improper, one would have indeed found the correct answer. So why should anyone care about the detail of this kind of improper integral? The answer becomes clear in the next example.

Example 5.2.8. Please compute

$$\int_{-1}^1 \frac{dx}{x^2}.$$

Answer. In this case, it may be best to present the wrong answer first:

$$\boxed{\int_{-1}^1 \frac{dx}{x^2} = \int_{-1}^1 x^{-2}\, dx = -\left(\frac{1}{x}\Big|_{-1}^1\right) = -2. \quad \textbf{This is False!}}$$

One of the ways that someone should immediately see that this answer cannot be correct is that the supposed answer is negative even though the integrand is positive on the entire interval. This of course is never possible.

Since the integrand is unbounded and undefined at $x = 0$, to find the correct answer, one must understand this integral as an improper integral:

$$\int_{-1}^1 \frac{dx}{x^2} := \lim_{\xi \to 0^-} \int_{-1}^\xi x^{-2}\, dx + \lim_{\xi \to 0^+} \int_\xi^1 x^{-2}\, dx = \lim_{\xi \to 0^-} \left(-\frac{1}{x}\Big|_{-1}^\xi\right) + \lim_{\xi \to 0^+} \left(-\frac{1}{x}\Big|_\xi^1\right).$$

Both of these limits *diverge*. This means that the integral in this example does not exist, no matter what one might find by ignoring the singularity of the integrand.

The importance of dealing with such improper integrals carefully with respect to the definition is difficult to overstate. The answer obtained incorrectly in Example 5.2.8 by ignoring that the integrand is unbounded was found only by in effect canceling out infinities—this is a dangerous idea. If such an integral represents, for example, force, the incorrect calculation given above might lead someone to think that everything is okay when in fact the force oscillates between very large positive and negative values. Such large oscillations could be very dangerous.

The final example of this section combines several of the issues discussed above and introduces one final integration technique: partial fractions.

Example 5.2.9. Please compute

$$\int_0^1 \frac{dx}{\sqrt[4]{x}(1 - \sqrt[4]{x})}.$$

Answer. Again, the first key observation in this case is that the integrand is singular, here at each of the two end points. This means that a variation on the definitions of improper integrals of the second kind is needed to deal with this situation. Before dealing with it, however, one must first notice that the integrand is not immediately integrable. This in fact is a messy example that shows both what is possible and what is not.

Notice that the denominator of the integrand is the product of two factors; one is zero at the lower limit of integration and the other is zero at the upper limit of integration. This suggests that the integral will need to be written as the sum of two integrals. To evaluate each of these integrals, the two factors need to be separated using partial fractions. From the partial fractions theorem, there exist constants A and B such that

$$\frac{1}{\sqrt[4]{x}(1 - \sqrt[4]{x})} = \frac{A}{\sqrt[4]{x}} + \frac{B}{1 - \sqrt[4]{x}}.$$

Multiplying each side by the original denominator, one finds that the constants here are $A = B = 1$. Thus,

$$\int_0^1 \frac{dx}{\sqrt[4]{x}(1 - \sqrt[4]{x})} = \int_0^1 \frac{dx}{\sqrt[4]{x}} + \int_0^1 \frac{dx}{1 - \sqrt[4]{x}}$$

provided each of the latter integrals exist. The first of these is improper at the lower limit of integration, so

$$\int_0^1 \frac{dx}{\sqrt[4]{x}} := \lim_{\delta \to 0^+} \int_\delta^1 x^{-1/4}\,dx = \lim_{\delta \to 0^+} \left(\frac{4}{3}x^{3/4}\right)\Big|_\delta^1 = \frac{4}{3} - \frac{4}{3}\lim_{\delta \to 0^+}\delta^{3/4} = \frac{4}{3}.$$

The second is improper at the upper limit of integration, so

$$\int_0^1 \frac{dx}{1 - \sqrt[4]{x}} := \lim_{\beta \to 1^-} \int_0^\beta \frac{dx}{1 - \sqrt[4]{x}}$$

$$= 4 \lim_{\beta \to 1^-} \int_{1-\sqrt[4]{\beta}}^1 \frac{(1-u)^3}{u} \, du = 4 \lim_{\beta \to 1^-} \int_{1-\sqrt[4]{\beta}}^1 \left(\frac{1}{u} - 3 + 3u - u^2 \right) du,$$

where $u = 1 - \sqrt[4]{x}$ and $x = (1 - u)^4$. All four of the terms in this final integrand can be evaluated directly, but only the first is really important here. It integrates to the natural logarithm, whose limit in turn diverges to negative infinity (because it is the lower limit) as $\beta \to 1^-$. Thus, even though the first integral converges to 4/3, the fact that the second integral diverges implies that the entire original integral in this example diverges to negative infinity (does not exist).

5.2.1 Partial fractions

Example 5.2.9 above is an example of the much broader topic of integration by partial fractions. Rather than state and prove the general theorem, in this case, let us consider a rather general example.

Example 5.2.10. Please find the partial fractions decomposition for

$$\frac{1}{(x - 1)^2(x + 2)}.$$

Answer. The key to a partial fractions decomposition is a theorem from the early 1700s. In the current example, this states that

$$\frac{1}{(x - 1)^2(x + 2)} = \frac{A}{(x - 1)^2} + \frac{B}{x - 1} + \frac{C}{x + 2}$$

for some coefficients A, B and C. To find these coefficients, we multiply the above equation by the denominator of the left hand side:

$$1 = A(x + 2) + B(x - 1)(x + 2) + C(x - 1)^2.$$

The values for A, B and C can now be found by multiplying out the right hand side and then equating the coefficients of the powers of x. Frequently, however, it is easier to simply set x equal to several helpful values. In this case, notice that setting $x = 1$ in the quadratic yields $1 = 3A$, while setting $x = -2$ yields $1 = 9C$. Then using these two values for A and C and setting $x = 0$ (or some other third value) yields $1 = 2A - 2B + C = 2/3 + 1/9 - 2B$, implying that $B = -1/9$. So the partial fractions decomposition for this rational expression is

$$\frac{1}{(x-1)^2(x+2)} = \frac{1}{3(x-1)^2} - \frac{1}{9(x-1)} + \frac{1}{9(x+2)}.$$

Of course this rather simple example does not cover all cases. If there is a numerator more complicated than 1, each term in the decomposition must be multiplied by this numerator, and long division may be required to reduce the degree of the numerator to be smaller than that of the denominator. Also the denominator may have irreducible quadratic factors (for example, $x^2 + x + 1$), meaning that terms of the form

$$\frac{Ax + B}{x^2 + x + 1}$$

(where the denominator must be replaced by whichever irreducible quadratic is actually present) may be needed in the decomposition.

Exercises 5, Group A

5.1. Please evaluate each of the following improper integrals when they exist or determine that the integral diverges and therefore does not exist.

(a) $\displaystyle\int_0^\infty e^{-x}\,dx$

(b) $\displaystyle\int_0^\infty te^{-t^2}\,dt$

(c) $\displaystyle\int_{-\infty}^\infty \frac{dx}{1+x^2}$

(d) $\displaystyle\int_0^1 \frac{dx}{\sqrt{1-x}}$

(e) $\displaystyle\int_{-2}^2 \frac{dx}{\sqrt[3]{x}}$

(f) $\displaystyle\int_{-2}^2 \frac{dx}{x^3}$

(g) $\displaystyle\int_0^{\pi/2} \tan\theta\,d\theta$

(h) $\displaystyle\int_0^\infty \frac{y}{e^y}\,dy$

(i) $\displaystyle\int_0^\infty \frac{dx}{1-x^2}$

(j) $\displaystyle\int_1^\infty \frac{dx}{x(1+x)}$

(k) $\displaystyle\int_e^\infty \frac{dx}{x(\ln x)^2}$

(l) $\displaystyle\int_1^\infty \frac{dx}{x(\ln x)^2}$

Answers. (a) Converges to 1. (c) Converges to π (the integral is the definition of arctan). (f) Diverges. (j) When computed carefully, converges to $\ln 2$ by partial fractions.

5.2. Please determine whether the following integrals converge or diverge by setting up the appropriate direct comparison. There is no need to evaluate any of these integrals, though for those that converge, this could be done numerically.

(a) $\displaystyle\int_{1}^{\infty} e^{-x^2}\,dx$

(c) $\displaystyle\int_{e}^{\infty} \frac{dx}{\sqrt{x}-1/x}$

(b) $\displaystyle\int_{-\infty}^{\infty} \frac{dx}{1+x^6}$

(d) $\displaystyle\int_{0}^{\infty} \frac{e^{-x}}{\sqrt{x}}\,dx$

Answers. (a) Converges by comparison with e^{-x}. (c) Diverges by comparison with $1/\sqrt{x}$.

5.3. Following the computation in Example 5.1.2, find the volume inside and the (outer) surface area of the solid of revolution obtained by revolving the curve $y = 1/x^2$ about the x-axis for $x \in [a, \infty)$, where $a > 0$. Notice that the surface area is now finite.

Answers. Volume: $\frac{\pi}{3a^3}$. Surface area: $\frac{2\pi}{a}$.

Exercises 5, Group B

5.1. Use a change of variables substitution in the integral definition of $\ln x$ (the natural logarithm function) to show that $\ln x^a = a \ln x$, where $a > 0$.

    Hint: Let $t = u^a$.

5.2. For $x, y > 0$, please show that each of the following identities is valid using the definition of the natural logarithm as an integral and the appropriate substitution.
(a) $\ln(xy) = \ln x + \ln y$
(b) $\ln(x/y) = \ln x - \ln y$
(c) $\ln(1/x) = -\ln x$
(d) $\ln x \to -\infty$ as $x \to 0^+$

    Hint: (a) $t = xu$.

5.3. Suppose that $q \in \mathbb{R}$. For which values of q does

$$\int_{0}^{1} x^q\,dx$$

exist as a finite number? What is the value of this integral when it exists?

Answer. For $q > -1$, the value is $1/(q + 1)$; for $q \le -1$, the integral does not exist.

5.4. In general, for $a, b \in \mathbb{R}$, it is not possible to evaluate

$$\int_{a}^{b} e^{-x^2}\,dx$$

exactly using the fundamental theorem of calculus because there is no (known, closed-form) antiderivative for the integrand in this integral. There is one important improper integral case, however, where an exact value can be found. Doing this requires the use of polar coordinates and integration in $\mathbb{R}^2$.

Consider the improper integral

$$\int_0^\infty e^{-x^2}\, dx$$

and in particular its square

$$\left(\int_0^\infty e^{-x^2}\, dx\right)^2 = \left(\int_0^\infty e^{-x^2}\, dx\right)\left(\int_0^\infty e^{-y^2}\, dy\right) = \int_0^\infty\int_0^\infty e^{-x^2-y^2}\, dxdy.$$

Writing the product of the two integrals as a single iterated integral (the rightmost integral) is possible because each integration variable is constant with respect to the other integration. This iterated integral now represents integration over the portion of $\mathbb{R}^2$ where both x and y are nonnegative. Notice that this portion of $\mathbb{R}^2$ can be described easily in polar coordinates:

$$\{(x,y) \mid x \geq 0, y \geq 0\} = \{(r,\theta) \mid r \geq 0, 0 \leq \theta \leq \pi/2\}.$$

So the iterated integral above can be written using polar coordinates, where in the multivariable coordinate change, $dxdy$ is replaced by $rdrd\theta$:

$$\int_0^\infty\int_0^\infty e^{-x^2-y^2}\, dxdy = \int_0^{\pi/2}\left(\int_0^\infty e^{-r^2} r\, dr\right)d\theta = \left(\int_0^{\pi/2} d\theta\right)\left(\int_0^\infty e^{-r^2} r\, dr\right).$$

The details of why the multivariable change of variables is what is claimed here are beyond the scope of this text (see Fehribach [2, Section 5.2] or any other text on multivariable calculus for details). The exercise here is just to evaluate each of the final two integrals above to find that

$$\int_0^\infty e^{-x^2}\, dx = \sqrt{\pi}/2.$$

5.5. Show that

$$\int_0^\infty x^n e^{-x}\, dx = n!$$

since this is the gamma function, $\Gamma(n+1)$.

Hint: Use induction starting with $n = 0$. For the inductive step, assume that the statement is true for $n = k$ and show that this implies the statement is true for $n = k + 1$. For

$$\int_0^\infty x^{k+1} e^{-x} \, dx,$$

use integration by parts and keep in mind that the inductive assumption is that

$$\int_0^\infty x^k e^{-x} \, dx = k!$$

Bibliography

[1] S. Abbott, *Understanding Analysis*, 2nd ed., Springer, New York, 2015.

[2] J. D. Fehribach, *Multivariable and Vector Calculus*, De Gruyter, Berlin, 2020.

[3] J. E. Graver, Listing the Positive Rationals, *Mathematics Magazine*, 94:1, pages 24–33, 2021. DOI: 10.1080/0025570X.2021.1843338.

[4] M. Rosenlicht, *Introduction to Analysis*, Dover, New York, 1986.

[5] W. Rudin, *Principles of Mathematical Analysis*, 3rd ed., McGraw-Hill, Inc., New York, 1976.

https://doi.org/10.1515/9783110768398-006

Index

absolute convergence 50, 63, 81
alternating harmonic series 49
alternating series 47, 50
alternating series test 47, 98
analytic function 86, 94
area under the curve 104
associativity 27, 65
axioms 26ff
– field 26, 27
– least upper bound 28–30
– order 27, 28

bounded 22
bounded above 28
bounded below 28

cancelation 26, 27
Cantor, Georg 26
Cantor diagonalization 35, 36
cardinality 33
Cauchy, Augustin-Louis VI, 21
Cauchy completeness 22, 29, 77
Cauchy convergence 21, 29, 77
Cauchy convergent 25
Cauchy ratio test 59, 82
Cauchy root test 61
Cauchy sequence 21, 29, 77
closure 27
commutativity 27, 63
completeness 22, 29
complex numbers, no order 28
conclusion 37
conditional convergence 50, 63
conjugate 6
continued roots 31, 40
contrapositive 37, 38
convergence 2, 12, 29, 41, 71, 72, 81, 112
– center 83
– endpoints 82
– interval 81, 83
– pointwise 71
– radius 81, 83
– uniform 72
convergence tests 46ff
– alternating series test 47
– comparison tests 51
– integral test 54

– n-th term test 46
– ratio test 59, 82
– root test 61
convergent sequence 2
converges absolutely 50
converges conditionally 50
converges pointwise 71
converges uniformly 72
converse 37, 38
countable 33
countably infinite 33

d'Alembert, Jean le Rond 59
d'Alembert Cauchy ratio test 59, 82
decreasing 30, 47
direct comparison test 51
Dirichlet, Gustav Lejeune 63
distributivity 27
divergence 2, 41, 81, 110, 112
divergence to infinity 8, 11, 14, 24

Eiffel Tower 38

field axioms 26, 27
fundamental contradiction in mathematics 63

Gabriel's horn 104
geometric series 43, 44, 81
greatest lower bound 28, 30

harmonic series 42
hypothesis 37

identities 27
if-then statements 36
implications 36
improper integral 55, 57, 101, 109
– first kind 55, 101
– second kind 109
increasing 30
indeterminate forms 11
– 0/0 8, 10
– 0^0 10
– 1^∞ 10
– ∞/∞ 4, 7, 8
– $\infty \cdot 0$ 10
– $\infty - \infty$ 11

https://doi.org/10.1515/9783110768398-007

www.ingramcontent.com/pod-product-compliance
Lightning Source LLC
Chambersburg PA
CBHW081544220326
41598CB00036B/6560